Butterworths Technical and Scientific Checkbooks

Physical Science 1 Checkbook

J O Bird
BSc(Hons), AFIMA, TEng(CEI), MITE

A J C May
BA, CEng, MIMechE, FITE, MBIM

Butterworth Scientific
London Boston Sydney Wellington Durban Toronto

All rights reserved. No part of this publication may be reproduced or transmitted in any form or by any means, including photocopying and recording without the written permission of the copyright holder, application for which should be addressed to the publishers. Such written permission must also be obtained before any part of this publication is stored in a retrieval system of any nature.

This book is sold subject to the Standard Conditions of Sale of Net Books and may not be resold in the UK below the net price given by the Publishers in their current price list.

First published 1982

© Butterworth & Co (Publishers) Ltd 1982

British Library Cataloguing in Publication Data

Bird, J.O.
   Physical science 1 checkbook.
   1. Science
   I. Title II. May, A.J.C.
   500.2'0246   Q160.2

ISBN 0-408-00681-1
ISBN 0-408-00628-5 Pbk

Typeset by Scribe Design, Gillingham, Kent
Printed in Scotland by Thomson Litho Ltd., East Kilbride

Contents

Preface vii

SECTION A — MATERIAL PROPERTIES AND STATICS

1 SI units and density 1
Main points 1
Worked problems 2
Further problems 4

2 The effects of forces on materials 7
Main points 7
Worked problems 9
Further problems 11

3 Atomic structure of matter 13
Main points 13
Worked problems 16
Further problems 18

4 Chemical reactions 20
Main points 20
Worked problems 22
Further problems 24

5 Forces in static equilibrium 27
Main points 27
Worked problems 30
Further problems 34

6 Pressure in fluids 38
Main points 38
Worked problems 39
Further problems 43

SECTION B — MOTION AND ENERGY

7 Speed and velocity 46
Main points 46
Worked problems 48
Further problems 51

8 Acceleration and force 55
Main points 55
Worked problems 56
Further problems 59

9 Friction 63
Main points 63
Worked problems 64
Further problems 66

10 Waves 69
  Main points 69
  Worked problems 72
  Further problems 72

11 Light rays 74
  Main points 74
  Worked problems 78
  Further problems 80

12 Work, energy and power 83
  Main points 83
  Worked problems 85
  Further problems 88

13 Heat energy 92
  Main points 92
  Worked problems 95
  Further problems 102

SECTION C – ELECTRICITY

14 Simple electric circuits 105
  Main points 105
  Worked problems 110
  Further problems 117

15 Resistance variation and electromagnetism 121
  Main points 121
  Worked problems 125
  Further problems 133

16 Chemical effects of electricity 138
  Main points 138
  Worked problems 141
  Further problems 145

Answers to multi-choice problems 148

Index 150

Note to Reader

As textbooks become more expensive, authors are often asked to reduce the number of worked and unworked problems, examples and case studies. This may reduce costs, but it can be at the expense of practical work which gives point to the theory.

Checkbooks if anything lean the other way. They let problem-solving establish and exemplify the theory contained in technician syllabuses. The Checkbook reader can gain *real* understanding through seeing problems solved and through solving problems himself.

Checkbooks do not supplant fuller textbooks, but rather supplement them with an alternative emphasis and an ample provision of worked and unworked problems. The brief outline of essential data—definitions, formulae, laws, regulations, codes of practice, standards, conventions, procedures, etc—will be a useful introduction to a course and a valuable aid to revision. Short-answer and multi-choice problems are a valuable feature of many Checkbooks, together with conventional problems and answers.

Checkbook authors are carefully selected. Most are experienced and successful technical writers; all are experts in their own subjects; but a more important qualification still is their ability to demonstrate and teach the solution of problems in their particular branch of technology, mathematics or science.

Authors, General Editors and Publishers are partners in this major low-priced series whose essence is captured by the Checkbook symbol of a question or problem 'checked' by a tick for correct solution.

Preface

This textbook of worked problems provides coverage of the Technician Education Council level I unit in Physical Science (syllabus U80/682, formerly U75/004). However it can be regarded as a basic textbook in Physical Science for a much wider range of courses, such as appropriate GCE O level or CSE courses.

The aims of the book are firstly, to develop in the student an understanding of the fundamental physical science concepts which will provide a common base for further studies in both science and technology by the student and, secondly, to give the student a basic knowledge of the concepts of density, chemical reactions, elasticity, temperature measurement, static equilibrium, pressure, straight line motion, waves, light rays, energy and electricity.

Each topic considered in the text is presented in a way that assumes in the reader little previous knowledge of that topic. This practical Physical Science book contains over 140 illustrations and approximately 170 detailed worked problems, followed by over 550 further problems with answers.

The authors would like to express their appreciation for the friendly co-operation and helpful advice given to them by the publishers. Thanks are due to Mrs. Elaine Woolley for the excellent typing of the manuscript.

Finally, the authors would like to add a word of thanks to their wives, Elizabeth and Juliet, for their patience, help and encouragement during the preparation of this book.

J O Bird
A J C May
Highbury College of Technology
Portsmouth

Butterworths Technical and Scientific Checkbooks

General Editors for Science, Engineering and Mathematics titles:
J.O. Bird and A.J.C. May, Highbury College of Technology, Portsmouth.

General Editor for Building, Civil Engineering, Surveying and Architectural titles:
Colin R. Bassett, lately of Guildford County College of Technology.

A comprehensive range of Checkbooks will be available to cover the major syllabus areas of the TEC, SCOTEC and similar examining authorities. A comprehensive list is given below and classified according to levels.

Level 1 (Red covers)
Mathematics
Physical Science
Physics
Construction Drawing
Construction Technology
Microelectronic Systems
Engineering Drawing
Workshop Processes & Materials

Level 2 (Blue covers)
Mathematics
Chemistry
Physics
Building Science and Materials
Construction Technology
Electrical & Electronic Applications
Electrical & Electronic Principles
Electronics
Microelectronic Systems
Engineering Drawing
Engineering Science
Manufacturing Technology
Digital Techniques
Motor Vehicle Science

Level 3 (Yellow covers)
Mathematics
Chemistry
Building Measurement
Construction Technology
Environmental Science
Electrical Principles
Electronics
Microelectronic Systems
Electrical Science
Mechanical Science
Engineering Mathematics & Science
Engineering Science
Engineering Design
Manufacturing Technology
Motor Vehicle Science
Light Current Applications

Level 4 (Green covers)
Mathematics
Building Law
Building Services & Equipment
Construction Technology
Construction Site Studies
Concrete Technology
Economics for the Construction Industry
Geotechnics
Engineering Instrumentation & Control

Level 5
Building Services & Equipment
Construction Technology
Manufacturing Technology

1 SI units and density

A. MAIN POINTS CONCERNED WITH SI UNITS AND DENSITY

1   The system of units used in engineering and science is the **Système Internationale d'Unités** (International system of units), usually abbreviated to SI units, and is based on the metric system. This was introduced in 1960 and is now adopted by the majority of countries as the official system of measurement.

2   Three of the basic SI units are listed below with their symbols:

Quantity	Unit
length, l	metre, m
mass, m	kilogram, kg
time, t	second, s

3   SI units may be made larger or smaller by using **prefixes** which denote multiplication or division by a particular amount. The four most common multiples, with their meaning, are listed below:

Prefix	Name	Meaning	
M	mega	multiply by 1 000 000	(i.e. $\times 10^6$)
k	kilo	multiply by 1000	(i.e. $\times 10^3$)
m	milli	divide by 1000	(i.e. $\times 10^{-3}$)
μ	micro	divide by 1 000 000	(i.e. $\times 10^{-6}$)

4   (i) **Length** is the distance between two points. The standard unit of length is the **metre**, although the **centimetre, cm, millimetre, mm** and **kilometre, km**, are often used.

   1 cm = 10 mm; 1 m = 100 cm = 1000 mm; 1 km = 1000 m.

   (ii) **Area** is a measure of the size or extent of a plane surface and is measured by multiplying a length by a length. If the lengths are in metres then the unit of area is the **square metre, m^2**.

   $1\ m^2 = 1\ m \times 1\ m = 100\ cm \times 100\ cm = 10\ 000\ cm^2$ or $10^4\ cm^2$
   $= 1000\ mm \times 1000\ mm = 1\ 000\ 000\ mm^2$ or $10^6\ mm^2$
   Conversely, $1\ cm^2 = 10^{-4}\ m^2$ and $1\ mm^2 = 10^{-6}\ m^2$

(iii) **Volume** is a measure of the space occupied by a solid and is measured by multiplying a length by a length by a length. If the lengths are in metres then the unit of volume is in **cubic metres, m³**.

$$1 \text{ m}^3 = 1 \text{ m} \times 1 \text{ m} \times 1 \text{ m} = 100 \text{ cm} \times 100 \text{ cm} \times 100 \text{ cm} = 10^6 \text{ cm}^3$$
$$= 1000 \text{ mm} \times 1000 \text{ mm} \times 1000 \text{ mm} = 10^9 \text{ mm}^3$$

Conversely, $1 \text{ cm}^3 = 10^{-6} \text{ m}^3$ and $1 \text{ mm}^3 = 10^{-9} \text{ m}^3$.

Another unit used to measure volume, particularly with liquids, is the litre, l, where $1 \text{ l} = 1000 \text{ cm}^3$.

(iv) **Mass** is the amount of matter in a body and is measured in **kilograms, kg**.

$1 \text{ kg} = 1000 \text{ g}$ (or conversely, $1 \text{ g} = 10^{-3} \text{ kg}$) and 1 tonne (t) = 1000 kg.

5 (i) **Density** is the mass per unit volume of a substance. The symbol used for density is ρ (Greek letter rho) and its units are kg/m³.

$$\text{Density} = \frac{\text{mass}}{\text{volume}}, \quad \text{i.e.,} \quad \boxed{\rho = \frac{m}{V}} \quad \text{or} \quad \boxed{m = \rho V} \quad \text{or} \quad \boxed{V = \frac{m}{\rho}}$$

where m is the mass in kg, V is the volume in m³ and ρ is the density in kg/m³.

(ii) Some typical values of densities include:

Aluminium	2700 kg/m³;	Steel	7800 kg/m³;
Cast iron	7000 kg/m³;	Petrol	700 kg/m³;
Cork	250 kg/m³;	Lead	11400 kg/m³;
Copper	8900 kg/m³;	Water	1000 kg/m³.

6 (i) The **relative density** of a substance is the ratio of the density of the substance to the density of water,

i.e. relative density = $\dfrac{\text{density of substance}}{\text{density of water}}$

Relative density has no units, since it is the ratio of two similar quantities.

(ii) Typical values of relative densities can be determined from para. 5 (since water has a density of 1000 kg/m³), and include:

Aluminium	2.7;	Steel	7.8;
Cast iron	7.0;	Petrol	0.7;
Cork	0.25;	Lead	11.4;
Copper	8.9;		

(iii) The relative density of a liquid (formerly called the 'specific gravity') may be measured using a **hydrometer**.

B. WORKED PROBLEMS ON SI UNITS AND DENSITY

Problem 1 Express (a) a length of 36 mm in metres, (b) 32 400 mm² in square metres, and (c) 8 540 000 mm³ in cubic metres.

(a) $1 \text{ m} = 10^3 \text{ mm}$ or $1 \text{ mm} = 10^{-3} \text{ m}$

Hence, $36 \text{ mm} = 36 \times 10^{-3} \text{ m} = \dfrac{36}{10^3} \text{ m} = \dfrac{36}{1000} \text{ m} = \mathbf{0.036 \text{ m}}$

(b) $1 \text{ m}^2 = 10^6 \text{ mm}^2$ or $1 \text{ mm}^2 = 10^{-6} \text{ m}^2$

Hence, $32\,400 \text{ mm}^2 = 32\,400 \times 10^{-6} \text{ m}^2 = \dfrac{32\,400}{10^6} = \mathbf{0.0324 \text{ m}^2}$

(c) $1 \text{ m}^3 = 10^9 \text{ mm}^3$ or $1 \text{ mm}^3 = 10^{-9} \text{ m}^3$

Hence, $8\,540\,000 \text{ mm}^3 = 8\,540\,000 \times 10^{-9} \text{ m}^3$

$= \dfrac{8\,540\,000}{10^9} \text{ m} = \mathbf{8.54 \times 10^{-3} \text{ m}^3 \text{ or } 0.00854 \text{ m}^3}$

Problem 2 Determine the area of a room 15 m long by 8 m wide in (a) m², (b) cm² and (c) mm².

(a) Area of room = $15 \text{ m} \times 8 \text{ m} = \mathbf{120 \text{ m}^2}$.
(b) $120 \text{ m}^2 = 120 \times 10^4 \text{ cm}^2$, since $1 \text{ m}^2 = 10^4 \text{ cm}^2$,
  $= \mathbf{1\,200\,000 \text{ cm}^2}$ or $\mathbf{1.2 \times 10^6 \text{ cm}^2}$.
(c) $120 \text{ m}^2 = 120 \times 10^6 \text{ mm}^2$, since $1 \text{ m}^2 = 10^6 \text{ mm}^2$,
  $= \mathbf{120\,000\,000 \text{ mm}^2}$ or $\mathbf{0.12 \times 10^9 \text{ mm}^2}$.

(Note, it is usual to express the power of 10 as a multiple of 3, i.e., $\times 10^3$ or $\times 10^6$ or $\times 10^{-9}$, and so on.)

Problem 3 A cube has sides each of length 50 mm. Determine the volume of the cube in cubic metres.

Volume of cube = $50 \text{ mm} \times 50 \text{ mm} \times 50 \text{ mm} = 125\,000 \text{ mm}^3$
$1 \text{ mm}^3 = 10^{-9} \text{ m}$, thus volume = $125\,000 \times 10^{-9} \text{ m}^3$.
$= \mathbf{0.125 \times 10^{-3} \text{ m}^3}$.

Problem 4 A container has a capacity of 2.5 litres. Calculate its volume in (a) m³, (b) mm³.

Since 1 litre = 1000 cm^3, 2.5 litres = $2.5 \times 1000 \text{ cm}^3 = 2500 \text{ cm}^3$.
(a) $2500 \text{ cm}^3 = 2500 \times 10^{-6} \text{ m}^3 = \mathbf{2.5 \times 10^{-3} \text{ m}^3}$ or $\mathbf{0.0025 \text{ m}^3}$.
(b) $2500 \text{ cm}^3 = 2500 \times 10^3 \text{ mm}^3 = \mathbf{2\,500\,000 \text{ mm}^3}$ or $\mathbf{2.5 \times 10^6 \text{ mm}^3}$.

Problem 5 Determine the density of 50 cm³ of copper if its mass is 445 g.

Volume = $50 \text{ cm}^3 = 50 \times 10^{-6} \text{ m}^3$; mass = $445 \text{ g} = 445 \times 10^{-3} \text{ kg}$.

Density = $\dfrac{\text{mass}}{\text{volume}} = \dfrac{445 \times 10^{-3} \text{ kg}}{50 \times 10^{-6} \text{ m}^3} = \dfrac{445}{50} \times 10^3 = \mathbf{8.9 \times 10^3 \text{ kg/m}^3}$ or $\mathbf{8900 \text{ kg/m}^3}$.

Problem 6 The density of aluminium is 2700 kg/m³. Calculate the mass of a block of aluminium which has a volume of 100 cm³.

Density, $\rho = 2700 \text{ kg/m}^3$; volume $V = 100 \text{ cm}^3 = 100 \times 10^{-6} \text{ m}^3$.

Since density = $\dfrac{\text{mass}}{\text{volume}}$, then mass = density $\times$ volume.

Hence mass = $\rho V = 2700 \text{ kg/m}^3 \times 100 \times 10^{-6} \text{ m}^3$

$= \dfrac{2700 \times 100}{10^6} \text{ kg} = \mathbf{0.270 \text{ kg}}$ or $\mathbf{270 \text{ g}}$.

Problem 7 Determine the volume, in litres, of 20 kg of paraffin oil of density 800 kg/m^3.

$$\text{Density} = \frac{\text{mass}}{\text{volume}} \text{ hence volume} = \frac{\text{mass}}{\text{density}}$$

Thus volume $= \frac{m}{\rho} = \frac{20 \text{ kg}}{800 \text{ kg/m}^3} = \frac{1}{40} \text{ m}^3 = \frac{1}{40} \times 10^6 \text{ cm}^3 = 25\,000 \text{ cm}^3$.

1 litre = 1000 cm³ hence $25\,000 \text{ cm}^3 = \frac{25\,000}{1000} = $ **25 litres**

Problem 8 Determine the relative density of a piece of steel of density 7850 kg/m^3. Take the density of water as 1000 kg/m^3.

$$\text{Relative density} = \frac{\text{density of steel}}{\text{density of water}} = \frac{7850}{1000} = 7.85$$

Problem 9 A piece of metal 200 mm long, 150 mm wide and 10 mm thick has a mass of 2700 g. What is the density of the metal?

Volume of metal = 200 mm × 150 mm × 10 mm = 300 000 mm³

$$= 3 \times 10^5 \text{ mm}^3 = \frac{3 \times 10^5}{10^9} \text{ m}^3 = 3 \times 10^{-4} \text{ m}^3$$

Mass = 2700 g = 2.7 kg.

$$\text{Density} = \frac{\text{mass}}{\text{volume}} = \frac{2.7 \text{ kg}}{3 \times 10^{-4} \text{ m}^3} = 0.9 \times 10^4 \text{ kg/m}^3 = \mathbf{9000 \text{ kg/m}^3}.$$

Problem 10 Cork has a relative density of 0.25. Calculate (a) the density of cork and (b) the volume in cubic centimetres of 50 g of cork. Take the density of water to be 1000 kg/m^3.

(a) Relative density $= \frac{\text{density of cork}}{\text{density of water}}$, from which,

density of cork = relative density × density of water,
i.e. density of cork, $\rho = 0.25 \times 1000 = \mathbf{250 \text{ kg/m}^3}$.

(b) Density $= \frac{\text{mass}}{\text{volume}}$, from which, volume $= \frac{\text{mass}}{\text{density}}$

Mass, $m = 50 \text{ g} = 50 \times 10^{-3} \text{ kg}$.

Hence volume, $V = \frac{m}{\rho} = \frac{50 \times 10^{-3}}{250 \text{ kg/m}^3} = \frac{0.05}{250} \text{ m}^3 = \frac{0.05}{250} \times 10^6 \text{ cm}^3 = \mathbf{200 \text{ cm}^3}$.

C. FURTHER PROBLEMS ON SI UNITS AND DENSITY

(a) SHORT ANSWER PROBLEMS

1   State the SI units for length, mass and time.

2   What is the meaning of the following prefixes? (a) M, (b) m, (c) μ, (d) k.

In *Problems 3 to 7*, complete the statements.

3 1 m = ...1000... mm; 1 km = ...1000... m

4 1 m² = ...10⁴... cm²; 1 cm² = ...100... mm²

5 1 l = ... cm³; 1 m³ = ...10⁹... mm³

6 1 kg = ...1000... g; 1 t = ...1000... kg

7 1 mm² = ...10⁻⁹... m²; 1 cm³ = ...10⁻⁶... m³

8 Define density.

9 What is meant by 'relative density'?

10 Relative density of liquids may be measured using a ....

(b) MULTI-CHOICE PROBLEMS (answers on page 148)

1 Which of the following statements is true?
   1000 mm³ is equivalent to (a) 1 m³; (b) 10^{-3} m³; (c) 10^{-6} m³; (d) 10^{-9} m³.

2 Which of the following statements is true?
   (a) 1 mm² = 10^{-4} m²;   (c) 1 mm³ = 10^{-6} m³;
   (b) 1 cm³ = 10^{-3} m³;   (d) 1 km² = 10^{10} cm².

3 Which of the following statements is false?
   1000 litres is equivalent to (a) 10^3 m³; (b) 10^6 cm³; (c) 10^9 mm³

4 Let mass = A, volume = B and density = C. Which of the following statements is false?
   (a) $A = BC$; (b) $C = \dfrac{A}{B}$; (c) $B = \dfrac{C}{A}$

5 The density of 100 cm³ of a material having a mass of 700 g is:
   (a) 70 000 kg/m³; (b) 7000 kg/m³; (c) 7 kg/m³; (d) 70 kg/m³.

(c) CONVENTIONAL PROBLEMS

1 Express (a) a length of 52 mm in metres, (b) 20 000 mm² in square metres, and (c) 10 000 000 mm³ in cubic metres.   [(a) 0.052 m; (b) 0.02 m²; (c) 0.01 m³]

2 A garage measures 5 m by 2.5 m. Determine the area in (a) m²; (b) mm².
   [(a) 12.5 m²; (b) 12.5 × 10⁶ mm²]

3 The height of the garage in *Problem 2* is 3 m. Determine the volume in
   (a) m³; (b) mm³.   [(a) 37.5 m³; (b) 37.5 × 10⁹ mm³]

4 A bottle contains 6.3 litres of liquid. Determine its volume in (a) m³; (b) cm³; (c) mm³.   [(a) 0.0063 m³; (b) 6300 cm³; (c) 6.3 × 10⁶ mm³]

5 Determine the density of 200 cm³ of lead which has a mass of 2280 g.
   [11 400 kg/m³]

6 The density of iron is 7500 kg/m³. If the volume of a piece of iron is 200 cm³, determine its mass.   [1.5 kg]

7 Determine the volume, in litres, of 14 kg of petrol of density 700 kg/m³.
   [20 litres]

8 The density of water is 1000 kg/m³. Determine the relative density of a piece of copper of density 8900 kg/m³.   [8.9]

9 A piece of metal 100 mm long, 80 mm wide and 20 mm thick has a mass of 1280 g. Determine the density of the metal. [8000 kg/m^3]

10 Some oil has a relative density of 0.80. Determine (a) the density of the oil, and (b) the volume of 2 kg of oil. Take the density of water as 1000 kg/m^3.
[(a) 800 kg/m^3; (b) 0.0025 m^3]

2 The effects of forces on materials

A. MAIN POINTS CONCERNED WITH THE EFFECTS OF FORCES ON MATERIALS

1 A **force** exerted on a body can cause a change in either the shape or the motion of the body. The unit of force is the **newton, N**.
2 No solid body is perfectly rigid and when forces are applied to it, changes in dimensions occur. Such changes are not always perceptible to the human eye since they are so small. For example, the span of a bridge will sag under the weight of a vehicle and a spanner will bend slightly when tightening a nut. It is important for engineers and designers to appreciate the effects of forces on materials, together with their mechanical properties.
3 The three main types of mechanical force that can act on a body are (i) tensile, (ii) compressive, and (iii) shear.

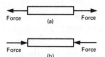
Fig 1

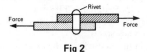
Fig 2

4 **Tensile force**
Tension is a force which tends to stretch a material, as shown in *Fig 1(a)*. Examples include:
  (i) the rope or cable of a crane carrying a load is in tension;
 (ii) rubber bands, when stretched, are in tension;
(iii) a bolt; when a nut is tightened, a bolt is under tension.
A tensile force, i.e., one producing tension, increases the length of the material on which it acts.
5 **Compressive force**
Compression is a force which tends to squeeze or crush a material, as shown in *Fig 1(b)*. Examples include:
  (i) a pillar supporting a bridge is in compression;
 (ii) the sole of a shoe is in compression;
(iii) the jib of a crane is in compression.
A compressive force, i.e. one producing compression, will decrease the length of the material on which it acts.

6 **Shear force**

Shear is a force which tends to slide one face of the material over an adjacent face. Examples include:
   (i) a rivet holding two plates together is in shear if a tensile force is applied between the plates (as shown in *Fig 2*);
   (ii) a guillotine cutting sheet metal, or garden shears, each provide a shear force;
   (iii) a horizontal beam is subject to shear force;
   (iv) transmission joints on cars are subject to shear forces.

   A shear force can cause a material to bend, slide or twist. (See *Problem 1*.)

7 **Elasticity** is the ability of a material to return to its original shape and size on the removal of external forces. If it does not return to the original shape, it is said to be **plastic**. Mild steel copper and rubber are examples of elastic materials; lead and plasticine are examples of plastic materials.

8 If a tensile force (i.e. a load) applied to a uniform bar of mild steel is gradually increased and the corresponding extension of the bar is measured, then a load-extension graph may be plotted as shown in *Fig 3*. Such an experiment is called a **tensile test**.

   Up to a certain point, marked A, the graph is a straight line, showing that in this region extension is proportional to the load. Point A is called the **elastic limit** and the straight line portion of the graph is called the **elastic stage**. In this region, when the load is removed the material will return to its original length. When the elastic limit is exceeded, the material becomes plastic and is altered in shape, this change in shape being permanent. Beyond point B shown in *Fig 3* the material will fracture.

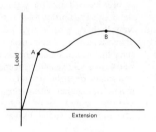

Fig 3

   The shape of load-extension graphs varies for different materials. (See *Problem 5*).

9 **Hooke's law** states:

   "Within the elastic limit, the extension of a material is proportional to the applied load."

   (See *Problems 2 to 4*.)

10 (i) **Ductility** is the ability of a material to be permanently stretched (i.e. drawn out to a small cross section by a tensile force). For ductile materials such as mild steel, copper and gold, large extensions can result before fracture occurs (see *Fig 3*).
    (ii) **Brittleness** is a lack of ductility. Brittle materials such as cast iron have virtually no plastic stage. The elastic stage is followed by immediate fracture (see *Problem 5(b)*). Other examples of brittle materials include glass, concrete, brick and ceramics.
    (iii) **Malleability** is the ability of a material to be permanently compressed without fracture (such as striking it with a hammer). Examples of malleable materials include lead, gold, putty and mild steel.

B. WORKED PROBLEMS ON THE EFFECTS OF FORCES ON MATERIALS

Problem 1 Fig 4(a) represents a crane and *Fig 4(b)* a transmission joint. State the types of forces acting, labelled A to F.

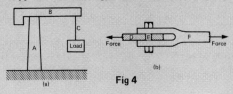

Fig 4

(a) *For the crane*
A, a supporting member, is in **compression**,
B, a horizontal beam, is in **shear**,
C, a rope, is in **tension**.

(b) *For the transmission joint*
D and F are in **tension**,
E, the rivet or bolt, is in **shear**.

Problem 2 A wire is stretched 2 mm by a force of 250 N. Determine the force that would stretch the wire 5 mm, assuming that the elastic limit is not exceeded.

Hooke's law states that extension x is proportional to force F, provided that the elastic limit is not exceeded, i.e., $x \propto F$ or $x = kF$ where k is a constant.

When $x = 2$ mm, $F = 250$ N, thus $2 = k(250)$

from which, constant, $k = \dfrac{2}{250} = \dfrac{1}{125}$

When $x = 5$ mm, then $5 = kF$, i.e., $5 = \left(\dfrac{1}{125}\right) F$

from which, force $F = 5(125) = 625$ N.

Thus to stretch the wire 5 mm a force of 625 N is required.

Problem 3 A force of 10 kN applied to a component produces an extension of 0.1 mm. Determine (a) the force needed to produce an extension of 0.12 mm, and (b) the extension when the applied force is 6 kN, assuming in each case that the elastic limit is not exceeded.

From Hooke's law, extension x is proportional to force F within the elastic limit, i.e., $x \propto F$ or $x = kF$, where k is a constant.
If a force of 10 kN produces an extension of 0.1 mm, then $0.1 = k(10)$

from which, constant, $k = \dfrac{0.1}{10} = 0.01$

(a) When extension $x = 0.12$ mm, then $0.12 = k(F)$, i.e., $0.12 = 0.01F$

from which, force $F = \dfrac{0.12}{0.01} = $ **12 kN**.

(b) When force $F = 6$ kN, then **extension,** $x = k(6) = (0.01)(6) = $ **0.06 mm**.

Problem 4 A tensile test is carried out on a mild steel specimen. The results are shown in the following table of values.

Load/kN	0	10	23	32
Extension/mm	0	0.023	0.053	0.074

Plot a graph of load against extension, and from the graph determine
(a) the load at an extension of 0.04 mm, and (b) the extension corresponding to a load of 28 kN.

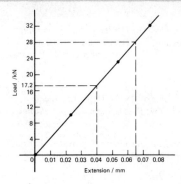

Fig 5

The load extension graph is shown in *Fig 5*.
From the graph:
(a) when the extension is 0.04 mm, the load is **17.2 kN**
(b) when the load is 28 kN, the extension is **0.065 mm**.

Problem 5 Sketch typical load extension curves for (a) an elastic non-metallic material, (b) a brittle material and (c) a ductile material. Give a typical example of each type of material.

(a) A typical load extension curve for an elastic non-metallic material is shown in *Fig 6(a)* and an example of such a material is **polythene**.

Fig 6

(b) A typical load extension curve for a brittle material is shown in *Fig 6(b)* and an example of such a material is **cast iron**.
(c) A typical load extension curve for a ductile material is shown in *Fig 6(c)* and an example of such a material is **mild steel**.

C. FURTHER PROBLEMS ON THE EFFECTS OF FORCES ON MATERIALS

(a) SHORT ANSWER PROBLEMS

1 Name three types of mechanical force that can act on a body.

2 State two practical examples of a tensile force.

3 State two practical examples of a compressive force.

4 State two practical examples of a shear force.

5 Define elasticity.

6 State Hooke's law.

7 What is the difference between a ductile and a brittle material?

8 Define malleability.

9 What is a 'tensile test'?

10 Sketch on the same axes a typical load extension graph for a ductile and a brittle material.

(b) MULTI-CHOICE PROBLEMS (answers on page 148)

1 A wire is stretched 3 mm by a force of 150 N. Assuming the elastic limit is not exceeded, the force that will stretch the wire 5 mm is:
(a) 150 N; (b) 250 N; (c) 90 N.

2 For the wire in *Problem 1*, the extension when the applied force is 450 N is:
(a) 1 mm; (b) 3 mm; (c) 9 mm.

3 The forces acting in a horizontal beam are in:
(a) tension; (b) compression; (c) shear.

4 A pillar supporting a bridge is an example of a force in:
(a) tension; (b) compression; (c) shear.

5 Which of the following statements is false?
(a) Elasticity is the ability of a material to return to its original dimensions after deformation by a load.
(b) Plasticity is the ability of a material to retain any deformation produced in it by a load.
(c) Ductility is the ability to be permanently stretched without fracturing.
(d) Brittleness is a lack of ductility and a brittle material has a long plastic stage.

(c) CONVENTIONAL PROBLEMS

1 Explain, using appropriate practical examples, the difference between tensile, compressive and shear forces.

2 (a) State Hooke's law.
   (b) A wire is stretched 1.5 mm by a force of 300 N. Determine the force that would stretch the wire 4 mm, assuming the elastic limit of the wire is not exceeded. [(b) 800 N]

3 A rubber band extends 50 mm when a force of 300 N is applied to it. Assuming the band is within the elastic limit, determine the extension produced by a force of 60 N. [10 mm]

4   A force of 25 kN applied to a piece of steel produces an extension of 2 mm. Assuming the elastic limit is not exceeded, determine (a) the force required to produce an extension of 3.5 mm, and (b) the extension when the applied force is 15 kN. [(a) 43.75 N; (b) 1.2 mm]

5   A coil spring 300 mm long when unloaded, extends to a length of 500 mm when a load of 40 N is applied. Determine the length of the spring when a load of 15 kN is applied. [375 mm]

6   A test to determine the load extension graph for a specimen of copper gave the following results:

Load/kN	8.5	15.0	23.5	30.0
Extension/mm	0.04	0.07	0.11	0.14

Plot the load extension graph, and from the graph determine (a) the load at an extension of 0.09 mm, and (b) the extension corresponding to a load of 12.0 N.
[(a) 19.1 kN; (b) 0.057 mm]

7   Define the following terms:
   (a) elasticity;
   (b) plasticity;
   (c) ductility;
   (d) brittleness
   (e) malleability

8   Sketch on the same axes typical load extension graphs for (a) a strong, ductile material, and (b) a brittle material.

3 Atomic structure of matter

A. MAIN POINTS CONCERNED WITH THE ATOMIC STRUCTURE OF MATTER

1 There is a very large number of different substances in existence, each substance containing one or more of a number of basic materials called elements. 'An **element** is a substance which cannot be separated into anything simpler by chemical means.' There are 92 naturally occurring elements and 13 others which have been artificially produced.

Some examples of common elements with their symbols are: Hydrogen H, Helium He, Carbon C, Nitrogen N, Oxygen O, Sodium Na, Magnesium Mg, Aluminium Al, Silicon Si, Phosphorus P, Sulphur S, Potassium K, Calcium Ca, Iron Fe, Nickel Ni, Copper Cu, Zinc Zn, Silver Ag, Tin Sn, Gold Au, Mercury Hg, Lead Pb and Uranium U.

2 Elements are made up of very small parts called atoms.
'An **atom** is the smallest part of an element which can take part in a chemical change and which retains the properties of the element.'

Each of the elements has a unique type of atom. An atom consists of electrons moving around a central nucleus containing protons and neutrons (see *Problem 1*).

3 When elements combine together, the atoms join to form a basic unit of a new substance. This independent group of atoms bonded together is called a molecule.
'A **molecule** is the smallest part of a substance which can have a separate stable existence.'

All molecules of the same substance are identical. Atoms and molecules are the **basic building blocks** from which matter is constructed.

4 When elements combine chemically their atoms interlink to form molecules of a new substance called a compound.
'A **compound** is a new substance containing two or more elements chemically combined so that their properties are changed.'

For example, the elements hydrogen and oxygen are quite unlike water, which is the compound they produce when chemically combined.

The components of a compound are in fixed proportion and are difficult to separate. Examples include:
(i) water H_2O, where 1 molecule is formed by 2 hydrogen atoms combining with 1 oxygen atom;
(ii) carbon dioxide, CO_2, where 1 molecule is formed by 1 carbon atom combining with 2 oxygen atoms;

(iii) sodium chloride NaCl (common salt), where 1 molecule is formed by 1 sodium atom combining with 1 chlorine atom, and

(iv) copper sulphate $CuSO_4$, where 1 molecule is formed by 1 copper atom, 1 sulphur atom and 4 oxygen atoms combining.

5. 'A **mixture** is a combination of substances which are not chemically joined together.'

    Mixtures have the same properties as their components. Also, the components of a mixture have no fixed proportion and are easy to separate. Examples include:
    (i) oil and water;
    (ii) sugar and salt;
    (iii) air, which is a mixture of oxygen, nitrogen, carbon dioxide and other gases;
    (iv) iron and sulphur;
    (v) sand and water.

    (See *Problems 2 and 3*)

6. 'A **solution** is a liquid in which other substances are dissolved.'

    A solution is a mixture from which the two constituents may not be separated by leaving it to stand or by filtration. For example, sugar dissolves in tea, salt dissolves in water and copper sulphate crystals dissolve in water leaving it a clear blue colour. The substance which is dissolved, which may be solid, liquid or gas, is called the **solute**, and the liquid in which it dissolves is called the **solvent**. Hence **solvent + solute = solution.**

    A solution has a clear appearance and remains unchanged with time.

7. 'A **suspension** is a mixture of a liquid and particles of a solid which do not dissolve in the liquid.'

    The solid may be separated from the liquid by leaving the suspension to stand or by filtration. Examples include:
    (i) sand in water;
    (ii) chalk in water;
    (iii) petrol and water.

    (See *Problem 4*)

8. (i) If a material dissolves in a liquid the material is said to be **soluble**. For example, sugar and salt are both soluble in water.

    (ii) If, at a particular temperature, sugar is continually added to water and the mixture stirred there comes a point when no more sugar can dissolve. Such a solution is called saturated. 'A solution is **saturated** if no more solute can be made to dissolve, with the temperature remaining constant.'

    (iii) '**Solubility** is a measure of the maximum amount of a solute which can be dissolved in 0.1 kg of a solvent, at a given temperature.'
    For example, the solubility of potassium chloride at 20°C is 34 g per 0.1 kg of water, or, its percentage solubility is 34%.

    (iv) The temperature of a mixture, the size of particles of the solute and the agitation of the mixture are factors which influence the solubility of a solid in a liquid.

    (See *Problems 5 and 6*)

9. A **crystal** is a regular, orderly arrangement of atoms or molecules forming a distinct pattern, i.e. an orderly packing of basic building blocks of matter. Most solids are crystalline in form and these include crystals such as common salt and sugar as well as the metals. Substances which are non-crystalline are called amorphous, examples including glass and wood. **Crystallization** is the process of isolating solids from solution in a crystalline form. This may be carried out by adding a solute to a solvent until saturation is reached, raising the temperature,

adding more solute and repeating the process until a fairly strong solution is obtained, and then allowing the solution to cool, when crystals will separate. There are several examples of crystalline form which occur naturally, examples including graphite, quartz, diamond and common salt.

10 Crystals can vary in size but always have a regular geometric shape with flat faces, straight edges and having specific angles between the sides. Two common shapes of crystals are shown in *Fig 1*. The angles between the faces of the common salt crystal (*Fig 1(a)*) are always 90° and those of a quartz crystal (*Fig 1(b)*) are always 60°. A particular material always produces exactly the same shape of crystal.

(a)

(b)

Fig 1

Fig 2 shows a crystal lattice of sodium chloride. This is always a cubic shaped crystal being made up of 4 sodium atoms and 4 chlorine atoms. The sodium chloride crystals then join together as shown.

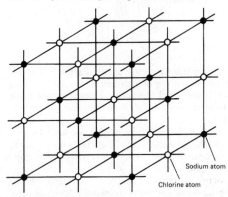

Fig 2

11 Metals are **polycrystalline** substances. This means that they are made up of a large number of crystals joined at the boundaries, the greater the number of boundaries the stronger the material.

12 Every metal, in the solid state, has its own crystal structure. To form an **alloy** different metals are mixed when molten, since in the molten state they do not have a crystal lattice. The molten solution is then left to cool and solidify. The solid formed is a mixture of different crystals and an alloy is thus referred to as a **solid solution.** Examples include:
  (i) brass, which is a combination of copper and zinc;
  (ii) steel, which is mainly a combination of iron and carbon;
  (iii) bronze, which is a combination of copper and tin.

Alloys are produced to enhance the properties of the metal, such as greater strength. For example, when a small proportion of nickel (say 2%–4%) is added to iron the strength of the material is greatly increased. By controlling the percentage of nickel added, materials having different specifications may be produced.

B. WORKED PROBLEMS ON THE ATOMIC STRUCTURE OF MATTER

Problem 1 Briefly describe a model depicting the structure of the atom.

In atomic theory, a model of an atom can be regarded as a miniature solar system. It consists of a central nucleus around which negatively charged particles called electrons orbit in certain fixed bands called shells. The nucleus contains positively charged particles called protons and particles having no electrical charge called neutrons.

An electron has a very small mass compared with protons and neutrons. An atom is electrically neutral, containing the same number of protons as electrons. The number of protons in an atom is called the atomic number of the element of which the atom is part. The arrangement of the elements in order of their atomic number is known as the periodic table.

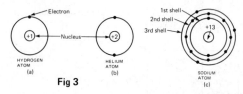

Fig 3

The simplest atom is hydrogen which has 1 electron orbiting the nucleus and 1 proton in the nucleus. The atomic number of hydrogen is thus 1. The hydrogen atom is shown diagrammatically in *Fig 3(a)*. Helium has 2 electrons orbiting the nucleus, both of them occupying the same shell at the same distance from the nucleus, as shown in *Fig 3(b)*.

The first shell of an atom can have up to 2 electrons only, the second shell can have up to 8 electrons only and the third shell up to 18 electrons only. Thus a sodium atom which has 13 electrons orbiting the nucleus is arranged as shown in *Fig 3(c)*.

Problem 2 State whether the following substances are elements, compounds or mixtures: (a) carbon, (b) salt, (c) mortar, (d) sugar, (e) copper.

(a) carbon is an **element**;
(b) salt, i.e. sodium chloride, is a **compound** of sodium and chlorine;
(c) mortar is a **mixture** of lime, sand and water;
(d) sugar is a **compound** of carbon, hydrogen and oxygen;
(e) copper is an **element**.

Problem 3 State four ways in which compounds can be distinguished from mixtures.

(i) The properties of a compound are different to its constituent components whereas a mixture has the same properties as its constituent components.
(ii) The components of a compound are in fixed proportion whereas the components of a mixture have no fixed proportion.
(iii) The atoms of a compound are joined whereas the atoms of a mixture are free.
(iv) When a compound is formed heat energy is produced or absorbed whereas when a mixture is formed little or no heat is produced or absorbed.

Problem 4 State whether the following mixtures are solutions or suspensions: (a) soda water, (b) chalk and water, (c) sea water, (d) petrol and water.

(a) soda water is a **solution** of carbon dioxide and water;
(b) chalk and water is a **suspension**, the chalk sinking to the bottom when left to stand;
(c) sea water is a **solution** of salt and water;
(d) petrol and water is a **suspension**, the petrol floating on the top of the water when left to stand.

Problem 5 Determine the solubility and the percentage solubility of common salt (sodium chloride) if, at a particular temperature, 180 g dissolves in 500 g of water.

The solubility is a measure of the maximum amount of sodium chloride which can be dissolved in 0.1 kg (i.e. 100 g) of water.

180 g of salt dissolves in 500 g of water.

Hence $\frac{180}{5}$ g of salt, i.e., 36 g, dissolves in 100 g of water.

Hence the solubility of sodium chloride is **36 g in 0.1 kg of water**.

Percentage solubility = $\frac{36}{100} \times 100\% = \mathbf{36\%}$.

Problem 6 List the common factors which influence the solubility of a solid in a liquid.

(i) Solubility is dependent on temperature. When solids dissolve in liquids, as the temperature is increased, in most cases the amount of solid that will go into solution also increases. (More sugar is dissolved in a cup of hot water than in the same amount of cold water.) There are exceptions to this for the solubility of common salt in water remains almost constant and the solubility of calcium hydroxide decreases as the temperature increases.
(ii) Solubility is obtained more quickly when small particles of a substance are added to a liquid than when the same amount is added in large particles. For example, sugar lumps take longer to dissolve in tea than does granulated sugar.
(iii) A solid dissolves in a liquid more quickly if the mixture is stirred or shaken, i.e., solubility depends on the speed of agitation.

Problem 7 Explain briefly how a metal may be (a) hardened, (b) annealed.

(a) A metal may be hardened by heating it to a high temperature then cooling it very quickly. This produces a large number of crystals and therefore many boundaries. The greater the number of crystal boundaries, the stronger is the metal.
(b) A metal is annealed by heating it to a high temperature and then allowing it to cool very slowly. This causes larger crystals, thus less boundaries and hence a softer metal.

C. FURTHER PROBLEMS ON THE ATOMIC STRUCTURE OF MATTER

(a) SHORT ANSWER PROBLEMS

1   What is an element?

2   Distinguish between atoms and molecules.

3   What is a compound?

4   Distinguish between compounds and mixtures.

5   Give three examples of an element.

6   Give three examples of a compound.

7   Give three examples of a mixture.

8   Define, and give one example of, a solution.

9   Define, and give one example of, a suspension.

10 Define solubility.

11 Define a saturated solution.

12 State three factors influencing the solubility of a solid in a liquid.

13 What is a crystal? Give three examples of those that occur naturally.

14 Briefly describe the process of crystallization from a solution.

15 What does polycrystalline mean?

16 How may a metallic alloy be formed?

17 Why are alloys formed?

18 Give three examples of metallic alloys.

(b) MULTI-CHOICE PROBLEMS (answers on page 148)

1   Which of the following statements is false?
    (a) The properties of a mixture are derived from the properties of its component parts.
    (b) Components of a compound are in fixed proportion.
    (c) In a mixture, the components may be present in any proportion.
    (d) The properties of compounds are related to the properties of their component parts.

2   Which of the following is a compound?
    (a) carbon; (b) silver; (c) salt; (d) ink.

3   Which of the following is a mixture?
    (a) air; (b) water; (c) lead; (d) salt.

4   Which of the following is a suspension?
    (a) soda water; (b) chalk and water; (c) lemonade; (d) sea water.

5   Which of the following is false?
    (a) carbon dioxide is a mixture;
    (b) sand and water is a suspension;
    (c) brass is an alloy;
    (d) common salt is a compound.

6 When sugar is completely dissolved in water the resulting clear liquid is termed a:
 (a) solvent; (b) solution; (c) solute; (d) suspension.

7 The solubility of potassium chloride is 34 g in 0.1 kg of water.
 The amount of water required to dissolve 510 g of potassium chloride is:
 (a) 173.4 g; (b) 1.5 kg; (c) 340 g; (d) 17.34 g.

8 Which of the following statements is false?
 (a) When two metals are combined to form an alloy, the strength of the resulting material is greater than that of either of the two original metals.
 (b) An alloy may be termed a solid solution.
 (c) In a solution the substance which is dissolved is called the solvent.
 (d) The atomic number of an element is given by the number of protons in the atom.

(c) CONVENTIONAL PROBLEMS

1 Explain the difference between a compound and a mixture and state three examples of each.

2 Describe with appropriate sketches, a model depicting the structure of the atom.

3 State whether the following are elements, compounds or mixtures:
 (a) town gas; (b) water; (c) oil and water; (d) aluminium.
 [(a) mixture; (b) compound; (c) mixture; (d) element]

4 Define solubility. How can the solubility of a solid in a liquid be influenced?

5 The solubility of sodium chloride is 0.036 kg in 0.1 kg of water.
 Determine the amount of water required to dissolve 432 g of sodium chloride.
 [1.2 kg]

6 Describe how an alloy may be formed and give four examples of alloys.

7 Define the terms (a) solute, (b) solvent, (c) solution, (d) suspension.

8 Explain, with the aid of a sketch, what is meant by a crystal, and give two examples of materials with a crystalline structure.

4 Chemical reactions

A. MAIN POINTS CONCERNED WITH CHEMICAL REACTIONS

1. A **chemical reaction** is an interaction between substances in which atoms are rearranged. A new substance is always produced in a chemical reaction.
2. **Air** is a mixture, and its composition by volume is approximately: nitrogen 78%, oxygen 21%, other gas (including carbon dioxide) 1%.
3. If a substance, such as powdered copper, of known mass, is heated in air, allowed to cool, and its mass remeasured, it is found that the substance has gained in mass. This is because the copper has absorbed oxygen from the air and changed into copper oxide. In addition, the proportion of oxygen in the air passed over the copper will decrease by the same amount as the gain in mass by the copper.
4. All materials require the presence of oxygen for burning to take place. Any substance burning in air will combine with the oxygen. This process is called **combustion**, and is an example of a chemical reaction between the burning substance and the oxygen in the air, the reaction producing heat. The chemical reaction is called **oxidation**.
5. An element reacting with oxygen produces a compound which contains only atoms of the original element and atoms of oxygen. Such compounds are called **oxides**. Examples of oxides include: copper oxide CuO, hydrogen oxide H_2O (i.e. water), and carbon dioxide CO_2.
6. **Rusting** of iron (and iron-based materials) is due to the formation on its surface of hydrated oxide of iron produced by a chemical reaction. Rusting of iron always requires the presence of oxygen and water.
7. Any iron or steel structure exposed to moisture is susceptible to rusting. This process, which cannot be reversed, can be dangerous since structures may be weakened by it. Rusting may be prevented by:
   (i) painting with water-resistant paint;
   (ii) galvanising the iron;
   (iii) plating the iron (see chapter 16, para. 7);
   (iv) an oil or grease film on the surface.
8. To represent a reaction a chemical shorthand is used. A symbol represents an element (such as H for hydrogen, O for oxygen, Cu for copper, Zn for zinc, and so on) and a formula represents a compound and gives the type and number of elements in the compound. For example, one molecule of sulphuric acid, H_2SO_4, contains 2 atoms of hydrogen, 1 atom of sulphur and 4 atoms of oxygen. Similarly,

a molecule of methane gas, CH_4, contains 1 atom of carbon and 4 atoms of hydrogen.

9 The rearrangement of atoms in a chemical reaction are shown by **chemical equations** using formulae and symbols.
For example:

(i) $S + O_2 = SO_2$

i.e. 1 molecule of sulphur, S, added to 1 molecule of oxygen, O_2, causes a reaction and produces 1 molecule of sulphur dioxide, SO_2.

(ii) $Zn + H_2SO_4 = ZnSO_4 + H_2$

i.e. 1 molecule of zinc, Zn, added to 1 molecule of sulphuric acid, H_2SO_4, causes a reaction and produces 1 molecule of zinc sulphate, $ZnSO_4$, and 1 molecule of hydrogen, H_2.

10 In a chemical equation:
(i) each element must have the same total number of atoms on each side of the equation. For example, in chemical equation (ii) of para. 9 each side of the equation has 1 zinc atom, 2 hydrogen atoms, 1 sulphur atom and 4 oxygen atoms.

(ii) a number written in front of a molecule multiplies all the atoms in that molecule. For example, the reaction described in para. 3 is:
$2Cu + O_2 = 2CuO$.

11 An **acid** is a compound containing hydrogen in which the hydrogen can be easily replaced by a metal. For example, in para. 9, it is shown that zinc reacts with sulphuric acid to give zinc sulphate and hydrogen.

An acid produces hydrogen ions H^+ in solution (an ion being a charged particle formed when atoms or molecules lose or gain electrons). Examples of acids include: Sulphuric acid, H_2SO_4, hydrochloric acid, HCl, and nitric acid, HNO_3.

12 A **base** is a substance which can neutralise an acid (i.e. remove the acidic properties of acids). An **alkali** is a soluble base. When in solution an alkali produces hydroxyl ions, OH^-. Examples of alkalis include: sodium hydroxide, NaOH (i.e., caustic soda), calcium hydroxide, $Ca(OH)_2$, ammonium hydroxide, NH_4OH and potassium hydroxide, KOH (caustic potash.)

13 A **salt** is the product of the neutralisation between an acid and a base, i.e. acid + base = salt + water. For example:

$HCl + NaOH = NaCl + H_2O$
$H_2SO_4 + 2KOH = K_2SO_4 + 2H_2O$
$H_2SO_4 + CuO = CuSO_4 + H_2O$

Examples of salts include: sodium chloride, NaCl (common salt), potassium sulphate, K_2SO_4, copper sulphate, $CuSO_4$, and calcium carbonate, $CaCO_3$ (limestone).

14 An **indicator** is a chemical substance, which when added to a solution, indicates the acidity or alkalinity of the solution by changing colour. Litmus is a simple two-colour indicator which turns red in the presence of acids and blue in the presence of alkalis. Two other examples of indicators are ethyl orange (red for acids, yellow for alkalis) and phenolphthalein (colourless for acids, pink for alkalis).

15 The **pH scale** (pH meaning 'the potency of hydrogen') represents, on a scale from 0 to 14, degrees of acidity and alkalinity. 0 is strongly acidic, 7 is neutral and

14 is strongly alkaline. Some average pH values include: concentrated hydrochloric acid HCl 1.0, lemon juice 3.0, milk 6.6, pure water 7.0, sea water 8.2, concentrated sodium hydroxide NaOH, 13.0.

B. WORKED PROBLEMS ON CHEMICAL REACTIONS

Problem 1 State the properties of oxygen and name some of its practical uses.

Oxygen is an odourless, colourless and tasteless element. It is slightly soluble in water (which is essential for fish), has a boiling point of $-183°C$ (i.e. 90 K), a freezing point of $-219°C$ (i.e. 54 K) and has approximately the same density as air. Oxygen is a strongly active chemical element and combines with many substances when they are heated.

Uses of oxygen include: chemical processing, metal cutting and welding processes to give a very hot flame when burnt with other gases, and for divers, mountaineers, firefighters using breathing apparatus and for medical use in hospitals.

Problem 2 Devise a simple experiment to demonstrate that water and air are necessary for rusting to occur.

Let a bright iron nail be placed in
(a) a test tube carrying tap water;
(b) a sealed test tube containing water from which air, and thus oxygen, has been removed by boiling;
(c) a dessicator containing dry air (a dessicator is an enclosed vessel containing a drying agent such as calcium chloride which absorbs all the water from the air);
(d) a test tube containing normal moist air.

After a period of time reddish-brown flakes of rust will appear on the iron nails in (a) and (d) but the nails in (b) and (c) will be unaffected. This simple experiment shows that water and air are necessary for rusting to occur.

Problem 3 Give an example of an oxide formation which is disadvantageous and state how it can be minimised.

An example of an oxide formation which is disadvantageous is the process of rusting. Rust will form on iron (and iron-based materials, such as steel) which are unprotected from moisture. Rusting can be minimised by covering the surface of the metal with a protective coating of paint, grease, oil, plastic or with a metal such as tin or zinc, this coating preventing air and water coming into contact with the iron.

Problem 4 State four examples of the damage done by rusting.

Examples of damage done by rusting may be found in:
(i) steel parts of a motor vehicle;
(ii) the hull of ships;
(iii) iron guttering;
(iv) bridge and similar structures.

Problem 5 Why should caustic soda be added very slowly to a solution?

Caustic soda (i.e. sodium hydroxide, NaOH) is mixed with cold water to produce a strong alkaline solution. During the mixing, the solution may well boil with the heat produced by the chemical reaction. For safety reasons, in order to limit the effect of the heat produced, the caustic soda is added very slowly to the water. The amount of heat produced is spread over a longer period and hence the heat produced is dissipated away from the solution. It is important that the water is never added to the alkali in solid form, since the alkali cannot readily dissipate the heat formed and pockets of steam from the water can cause a minor explosion.

Problem 6 State the essential differences between a chemical change and a physical change. Give two examples of each type of change.

(i) A chemical change always produces at least one new substance, whereas with a physical change no new substances are formed.
(ii) In a chemical change, energy in the form of heat is frequently liberated or sometimes absorbed whereas with a physical change little energy change is involved.
(iii) A chemical change is very difficult to reverse whereas a physical change is normally simple to reverse.
(iv) In a chemical change the masses of the individual new compounds formed are usually different from the masses of the original chemicals; in a physical change each substance retains its mass throughout.

The rusting of iron and the burning of coal are examples of chemical changes.
The magnetising of a piece of steel and the mixing of sand and pebbles are examples of physical changes.

Problem 7
(a) Write an equation to represent the reaction between magnesium (Mg) and sulphuric acid (H_2SO_4) to form magnesium sulphate ($MgSO_4$) and hydrogen (H_2).
(b) A neutralisation reaction occurs when copper oxide CuO is added to sulphuric acid. Copper sulphate and water is produced. Write down the chemical equation and state which is the base, the acid and the salt.

(a) $Mg + H_2SO_4 = MgSO_4 + H_2$

(b) $CuO + H_2SO_4 = CuSO_4 + H_2O$

i.e. base + acid = salt + water.

Problem 8 State four properties of acids and four properties of alkalis.

Acids have the following properties:
(i) Almost all acids react with carbonates and bicarbonates, (a carbonate being a compound containing carbon and oxygen—an example being sodium carbonate, i.e. washing soda).
(ii) Dilute acids have a sour taste. Examples include citric acid (lemons), acetic acid (vinegar) and lactic acid (sour milk).

(iii) Acid solutions turn litmus paper red, methyl orange red and phenolphthalein colourless.
(iv) Most acids react with the higher elements in the electrochemical series (see chapter 16) and hydrogen is released.

Alkalis have the following properties:
(i) Alkalis neutralise acids to form a salt and water only.
(ii) Alkalis have little effect on metals.
(iii) Alkalis turn litmus paper blue, methyl orange yellow and phenolphthalein pink.
(iv) Alkalis are slippery when handled. Strong alkalis are good solvents for certain oils and greases.

C. FURTHER PROBLEMS ON CHEMICAL REACTIONS

(a) SHORT ANSWER PROBLEMS

1 What is a chemical reaction?

2 Air is a mixture, mainly of .... and ....

3 What happens to the mass of some copper when it is heated in air?

4 What is an oxide?

5 What is rust?

6 Rusting of iron requires the presence of .... and ....

7 State how rust may be prevented.

8 What is a chemical equation?

9 Define an acid.

10 Give three examples of acids.

11 Define a base.

12 Define an alkali.

13 Give three examples of alkalis.

14 Define a salt.

15 Describe how acidity/alkalinity may be determined by means of indicators.

(b) MULTI-CHOICE PROBLEMS (answers on page 148)

1 Rust occurring on iron railings is an example of:
(a) a mixture; (b) a solution; (c) an element; (d) a compound.

2 The chemical formula for sulphuric acid:
(a) H_2SO_2; (b) H_2SO_3; (c) H_2SO_4; (d) HSO_4.

3 Which of the following statements is true?
Acids
(a) turn litmus paper blue;
(b) are soluble in water;

(c) make methyl orange indicator yellow;
(d) have a pH value greater than 7.

4   Which of the following statements is false?
An alkali:
(a) is a soluble base;
(b) when in solution, produces hydroxyl ions;
(c) has a pH value less than 7;
(d) turns litmus paper blue.

5   1 molecule of nitric acid HNO_3 contains the following number of atoms. Which is true?
(a) 3 hydrogen, 1 nitrogen and 1 oxygen;
(b) 1 hydrogen, 1 nitrogen and 3 oxygen;
(c) 1 hydrogen, 3 nitrogen and 1 oxygen;
(d) 1 hydrogen, 1 nitrogen and 1 oxygen.

6   Which of the following statements is false?
(a) Hydrogen oxide is the chemical name for water;
(b) Calcium hydroxide is an example of an alkali;
(c) Calcium carbonate is an example of a salt;
(d) Sodium hydroxide is an example of an acid.

7   Which of the following chemical equations is incorrect?
(a) $Na + H_2O = NaOH + H_2$;
(b) $Cu + H_2SO_4 = CuSO_4 + H_2$;
(c) $2K + 2H_2O = 2KOH + H_2$;
(d) $CaO + H_2O = Ca(OH)_2$;

8   Which of the following chemical equations is incorrect?
(a) $Pb + H_2SO_4 = PbSO_4 + H_2$;
(b) $2Mg + O_2 = 2MgO$;
(c) $HNO_3 + NaOH = NaNO_3 + H_2O$;
(d) $2H_2O = H_2 + O_2$.

(c) CONVENTIONAL PROBLEMS

1   Explain the differences between the terms acid, base and salt and give one example of each.

2   Explain the difference between a chemical change and a physical change and state three examples of each.

3   List the main characteristics of oxygen and state practical uses of it.

4   (a) What is meant by oxidation?
(b) Explain briefly the process of rusting and state how it may be minimised.

5   Write down the chemical equation for the reaction that occurs when potassium hydroxide, KOH, is added to sulphuric acid, H_2SO_4.
$$[2KOH + H_2SO_4 = K_2SO_4 + 2H_2O]$$

6   State the chemical symbols for (a) copper-oxide, (b) hydrochloric acid, (c) sodium hydroxide, (d) nitric acid, (e) calcium hydroxide, (f) zinc sulphate.
$$[(a)\ CuO;\ (b)\ HCl;\ (c)\ NaOH;\ (d)\ HNO_3;\ (e)\ Ca(OH)_2;\ (f)\ ZnSO_4]$$

7 (i) State the effects of acids and alkalis on (a) litmus paper, (b) methyl orange indicator and (c) phenolphthalein indicator.
   (ii) What is meant by the pH scale?

8 Some copper is placed in sulphuric acid. State the chemical equation representing the reaction. $[Cu + H_2SO_4 = CuSO_4 + H_2]$

9 State the properties of acids and alkalis. Give three examples of acids and three examples of alkalis.

5 Forces in static equilibrium

A. MAIN POINTS CONCERNING FORCES IN STATIC EQUILIBRIUM

1. Quantities used in engineering and science can be divided into two groups:
   (a) **Scalar quantities** have a size or magnitude only and need no other information to specify them. Thus, 10 centimetres, 50 seconds, 7 litres and 3 kilograms are all examples of scalar quantities.
   (b) **Vector quantities** have both a size or magnitude and a direction, called the line of action of the quantity. Thus, a velocity of 50 kilometres per hour due east, on acceleration of 9.8 metres per second squared vertically downwards and a force of 15 newtons at an angle of 30 degrees are all examples of vector quantities.

2. The **centre of gravity** of an object is a point where the resultant gravitational force acting on the body may be taken to act. For objects of uniform thickness lying in a horizontal plane, the centre of gravity is vertically in line with the point of balance of the object. For a thin uniform rod the point of balance and hence the centre of gravity is halfway along the rod as shown in *Fig 1(a)*.

   A thin flat sheet of a material of uniform thickness is called a **lamina** and the centre of gravity of a rectangular lamina lies at the point of intersection of its diagonals, as shown in *Fig 1(b)*. The centre of gravity of a circular lamina is at the centre of the circle, as shown in *Fig 1(c)*.

(a)

(b)

(c)

Fig 1

3. An object is in **equilibrium** when the forces acting on the object are such that there is no tendency for the object to move. The state of equilibrium of an object can be divided into three groups.
   (i) If an object is in **stable equilibrium** and it is slightly disturbed by pushing or pulling, (i.e., a disturbing force is applied), the centre of gravity is raised and when the disturbing force is removed, the object returns to its original position. Thus a ball bearing in a hemispherical cup is in stable equilibrium, as shown in *Fig 2(a)*.

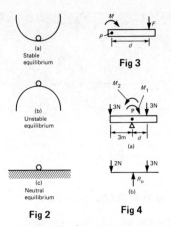

Fig 2

Fig 3

Fig 4

(ii) An object is in **unstable equilibrium** if, when a disturbing force is applied, the centre of gravity is lowered and the object moves away from its original position. Thus, a ball bearing balanced on top of a hemispherical cup is in unstable equilibrium, as shown in *Fig 2(b)*.

(iii) When an object in **neutral equilibrium** has a disturbing force applied, the centre of gravity remains at the same height and the object does not move when the disturbing force is removed. Thus, a ball bearing on a flat horizontal surface is in neutral equilibrium, as shown in *Fig 2(c)*.

4 When using a spanner to tighten a nut, a force tends to turn the nut in a clockwise direction. This turning effect of a force is called the **moment of a force** or more briefly, just a **moment**. The size of the moment acting on the nut depends on two factors:

(a) the size of the force acting at right angles to the shank of the spanner, and
(b) the perpendicular distance between the point of application of the force and the centre of the nut.

In general, with reference to *Fig 3*: moment of a force acting at a point P
= force × perpendicular distance between the line of action of the force and P.

i.e. $M = F \times d$

The unit of a moment is the newton metre, (N m). Thus, if force F in *Fig 3* is 7 N and distance d is 3 m, then the moment P is 7(N) × 3(m), i.e., 21 N m.
(See *Problems 1 and 2*)

5 If more than one force is acting on an object and the forces do not act at a point, then the turning effect of the forces, that is, the moment of the forces, must be considered. For equilibrium, the total moment of the forces acting on an object must be zero. For the total moment to be zero:
'the sum of the clockwise moments about any point must be equal to the sum of the anticlockwise moments about that point'.

This statement is known as the **principle of moments**. For the system of forces shown in *Fig 4*, taking moments about point P gives a moment, M_1, of 3(N) × d(m), i.e., $M_1 = 3d$ (N m), this moment tending to turn the system in a clockwise direction. The moment, M_2, is 2(N) × 3(m), i.e., $M_2 = 6$ (N m), tending to turn the system in an anticlockwise direction. Applying the principle of moments, for the system of forces to be in equilibrium:

clockwise moment = anticlockwise moment
hence, $3d = 6$
from which, $d = \frac{6}{3} = 2$ m.

In addition, for the system to be in equilibrium, the sum of the forces acting vertically downwards must be equal to the sum of the forces acting vertically

upwards, hence the force acting on the beam support, $R_p = (2+3)\text{N} = 5\text{ N}$, as shown in *Fig 4(b)*.
(See *Problems 3 to 6*)

6 As stated in para. 1, force is a vector quantity and thus has both a magnitude and a direction. A vector can be represented graphically by a line drawn to scale in the direction of the line of action of the force. Vector quantities may be shown by using bold, lower case letters, thus **ab** in *Fig 5* represents a force of 5 newtons acting in a direction due east.

For two forces acting at a point, there are three possibilities.

(a) For forces acting in the same direction and having the same line of action, the single force having the same effect as both of the forces, called the **resultant force** or just the **resultant**, is the arithmetic sum of the separate forces. Forces of F_1 and F_2 acting at point P, as shown in *Fig 6(a)* have

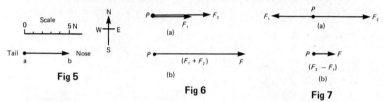

Fig 5

Fig 6

Fig 7

exactly the same effect on point P as force F shown in *Fig 6(b)*, where $F = F_1 + F_2$ and acts in the same direction as F_1 and F_2. Thus, F is the resultant of F_1 and F_2.
(See *Problem 7*)

(b) For forces acting in opposite directions along the same line of action, the resultant force is the arithmetic difference between the two forces. Forces of F_1 and F_2 acting at point P as shown in *Fig 7(a)*, have exactly the same effect on point P as force F shown in *Fig 7(b)*, where $F = F_2 - F_1$ and acts in the direction of F_2, since F_2 is greater than F_1. Thus F is the resultant of F_1 and F_2.
(See *Problem 7*)

(c) When two forces do not have the same line of action, the magnitude and direction of the resultant force may be found by a procedure called **vector addition** of forces. There are two graphical methods of performing vector addition, known as the triangle of forces method and the parallelogram of forces method.

The triangle of forces method:
  (i) Draw a vector representing one of the forces, using an appropriate scale and in the direction of its line of action.
  (ii) From the **nose** of this vector and using the same scale, draw a vector representing the second force in the direction of its line of action.
  (iii) The resultant vector is represented in both magnitude and direction by the vector drawn from the **tail** of the first vector to the nose of the second vector.

(See *Problems 8 and 9*)

The parallelogram of forces method:
  (iv) Draw a vector representing one of the forces, using an appropriate scale and in the direction of its line of action.
  (v) From the **tail** of this vector and using the same scale draw a vector representing the second force in the direction of its line of action.

(vi) Complete the parallelogram using the two vectors drawn in (iv) and (v) as two sides of the parallelogram.
(vii) The resultant force is represented in both magnitude and direction by the vector corresponding to the diagonal of the parallelogram drawn from the tail of the vectors in (iv) and (v).
(See *Problem 10*)

B. WORKED PROBLEMS ON FORCES IN STATIC EQUILIBRIUM

Problem 1 A force of 15 N is applied to a spanner at an effective length of 140 mm from the centre of a nut. Calculate (a) the moment of the force applied to the nut and (b) the magnitude of the force required to produce the same moment if the effective length is reduced to 100 mm.

From para. 4, $M = F \times d$, where M is the turning moment, F is the force applied at right angles to the spanner and d is the effective length between the force and the centre of the nut. Thus, with reference to *Fig 8(a)*:

(a) turning moment,
$$M = 15 \text{ N} \times 140 \text{ mm}$$
$$= 2100 \text{ N mm}$$
$$= 2100 \text{ N mm} \times \frac{1 \text{ m}}{1000 \text{ mm}}$$
$$= 2.1 \text{ N m}.$$

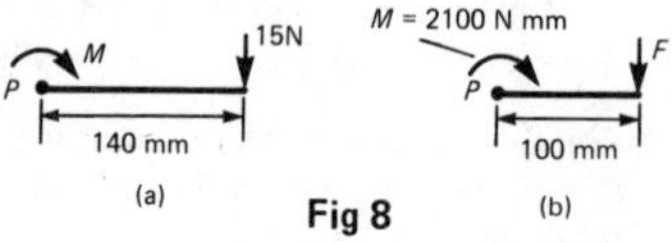

Fig 8

(b) The turning moment, M, is 2100 N mm and the effective length d becomes 100 mm (see *Fig 8(b)*). Applying $M = F \times d$ gives

$$2100 \text{ N mm} = F \times 100 \text{ mm}$$

from which, force $F = \dfrac{2100 \text{ N mm}}{100 \text{ mm}} = \mathbf{21 \text{ N}}.$

Problem 2 A moment of 25 N m is required to operate a lifting jack. Determine the effective length of the handle of the jack if the force applied to it is (a) 125 N and (b) 0.4 kN.

From para. 4, moment $M = F \times d$, where F is the force applied at right angles to the handle and d is the effective length of the handle. Thus:

(a) $25 \text{ N m} = 125 \text{ N} \times d$

from which, effective length, $d = \dfrac{25 \text{ N m}}{125 \text{ N}} = \dfrac{1}{5} \text{ m}$

$$= \frac{1}{5} \text{ m} \times \frac{1000 \text{ mm}}{1 \text{ m}} = \frac{1000}{5} \text{ mm}$$

$$= \mathbf{200 \text{ mm}}.$$

(b) The turning moment, M, is 25 N m and the force F becomes 0.4 kN, i.e. 400 N. Since $M = F \times d$, then $25 \text{ N m} = 400 \text{ N} \times d$.

Thus, the effective length, $d = \dfrac{25 \text{ N m}}{400 \text{ N}} = \dfrac{1}{16} \text{ m}$

$$= \frac{1}{16} \text{ m} \times \frac{1000 \text{ mm}}{1 \text{ m}} = \mathbf{62.5 \text{ mm}}.$$

Problem 3 A system of force is as shown
in *Fig 9*. (a) If the system is in equilibrium
find the distance d. (b) If the point of
application of the 5 N force is moved to
point P, distance 200 mm from the
support, find the new value of F to
replace the 5 N force for the system to
be in equilibrium.

Fig 9

From para. 5:
(a) the clockwise moment M_1 is due to a force of 7 N acting at a distance d from the support, (called the **fulcrum**), i.e.,

$M_1 = 7 \text{ N} \times d$

The anticlockwise moment M_2 is due to a force of 5 N acting at a distance of 140 mm from the fulcrum, i.e.

$M_2 = 5 \text{ N} \times 140 \text{ mm}.$

Applying the principle of moments, for the system to be in equilibrium:

clockwise moment = anticlockwise moment
i.e. $7 \text{ N} \times d$ = 5×140 N mm
Hence, distance $d = \dfrac{5 \times 140 \text{ N mm}}{7 \text{ N}} = 100$ mm

(b) When the 5 N force is replaced by force F at a distance of 200 mm from the fulcrum, the new value of the anticlockwise moment is $F \times 200$. For the system to be in equilibrium:

clockwise moment = anticlockwise moment
i.e. (7×100) N mm = $F \times 200$ mm
Hence, new value of force, $F = \dfrac{700 \text{ N mm}}{200 \text{ mm}} = 3.5$ N.

Problem 4 A beam is supported at its
centre on a fulcrum and forces act as
shown in *Fig 10*. Calculate (a) force F
for the beam to be in equilibrium and
(b) the new position of the 23 N force
when F is decreased to 21 N, if equilibrium is to be maintained.

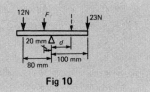

Fig 10

From para. 5:
(a) the clockwise moment, M_1, is due to the 23 N force acting at a distance of 100 mm from the fulcrum, i.e.,

$M_1 = 23 \times 100 = 2300$ N mm

There are two forces giving the anticlockwise moment M_2. One is the force F acting at a distance of 20 mm from the fulcrum and the other a force of 12 N acting at a distance of 80 mm.

Thus $M_2 = (F \times 20 + 12 \times 80)$ N mm

Applying the principle of moments:

clockwise moment = anticlockwise moments
i.e. 2300 = $F \times 20 + 12 \times 80$
Hence, $F \times 20$ = 2300−960
i.e. force F = $\dfrac{1340}{20}$ = **67 N**.

(b) The clockwise moment is now due to a force of 23 N acting at a distance of, say, d, from the fulcrum. Since the value of F is decreased to 21 N, the anticlockwise moment is $(21 \times 20 + 12 \times 80)$ N mm. Applying the principle of moments:

$23 \times d$ = $21 \times 20 + 12 \times 80$
i.e. distance $d = \dfrac{420+960}{23} = \dfrac{1380}{23}$
= **60 mm**.

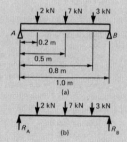

Problem 5 A beam is loaded as shown in *Fig 11*. Determine (a) the force acting on the beam support at B and (b) the force acting on the beam support at A, neglecting the mass of the beam.

Fig 11

A beam supported as shown in *Fig 11* is called a **simply supported beam**.

(a) Taking moments about point A and applying the principle of moments gives:

clockwise moments = anticlockwise moments

$(2 \times 0.2 + 7 \times 0.5 + 3 \times 0.8)$ kN m = $R_B \times 1.0$ m, where R_B is the force supporting the beam at B as shown in *Fig 11(b)*.
Thus $(0.4 + 3.5 + 2.4)$ kN m = $R_B \times 1.0$ m

i.e. $R_B = \dfrac{6.3 \text{ kN m}}{1.0 \text{ m}} = \mathbf{6.3 \text{ kN}}$

(b) For the beam to be in equilibrium, the forces acting upwards must be equal to the forces acting downwards, thus

$R_A + R_B = (2 + 7 + 3)$ kN

But $R_B = 6.3$ kN, thus $R_A = 12 - 6.3 = \mathbf{5.7 \text{ kN}}$.

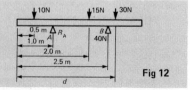

Problem 6 For the beam shown in *Fig 12*, calculate (a) the force acting on support A and (b) distance d, neglecting any forces arising from the mass of the beam.

Fig 12

(a) From para. 5, (the forces acting in an upward direction) = (the forces acting in a downward direction)

Hence, $(R_A + 40)$ N $= (10 + 15 + 30)$ N

$R_A = 10 + 15 + 30 - 40 = \mathbf{15\ N}.$

(b) Taking moments about the left hand end of the beam and applying the principle of moments gives:

clockwise moments = anticlockwise moments

$(10 \times 0.5 + 15 \times 2.0)$ N m $+ 30$ N $\times d = (15 \times 1.0 + 40 \times 2.5)$ N m

i.e., 35 N m $+ 30$ N $\times d = 115$ N m,

from which, distance, $d = \dfrac{(115-35)\ \text{N m}}{30\ \text{N}} = 2\dfrac{2}{3}\text{m}$

Problem 7 Determine the resultant of forces of 5 kN and 8 kN, (a) acting in the same direction and having the same line of action, and (b) acting in opposite directions but having the same line of action.

(a) The vector diagram of the two forces acting in the same direction is shown in *Fig 13(a)*, which assumes that the line of action is horizontal although, since it is not specified could be in any direction. From para. 6(a), the resultant force F is given by:

$F = F_1 + F_2$

i.e. $F = (5 + 8)$ kN $= \mathbf{13\ kN}$ in the direction of the original forces.

(b) The vector diagram for the two forces acting in opposite directions is shown in *Fig 13(b)*, again assuming that the line of action is in a horizontal direction. From para. 6(b), the resultant force F is given by:

$F = F_2 - F_1$

i.e. $F = (8-5)$ kN $= \mathbf{3\ kN}$ in the direction of the 8 kN force.

Problem 8 Determine the magnitude and direction of the resultant of a force of 15 N acting horizontally to the right and a force of 20 N, inclined at an angle of 60° to the 15 N force. Use the triangle of forces method.

Using the procedure given in para. 6(c) and with reference to *Fig 14*.

(i) **ab** is drawn 15 units long horizontally.
(ii) from b, **bc** is drawn 20 units long, inclined at an angle of 60° to **ab**. [Note, in angular measure, an angle of 60° from **ab** means 60° in an anticlockwise direction.]
(iii) By measurement, the resultant **ac** is 30.5 units long inclined at an angle of 35° to **ab**.

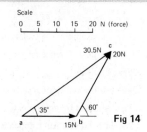

Fig 14

That is, the resultant force is **30.5 N** inclined at an angle of **35°** to the 15 N force.

Problem 9 Find the magnitude and direction of the two forces given, using the triangle of forces method.
First force: 1.5 kN acting at an angle of 30°
Second force: 3.7 kN acting at an angle of −45°.

From the procedure given in para. 6(c) and with reference to *Fig 15*:
(i) **ab** is drawn at an angle of 30° and 1.5 units in length;
(ii) From b, **bc** is drawn at an angle of −45° and 3.7 units in length; [Note, an angle of −45° means a clockwise rotation of 45° from a line drawn horizontally to the right],
(iii) By measurement, the resultant **ac** is 4.3 units long at an angle of −25°.

That is, the resultant force is **4.3 kN** at an angle of **−25°**.

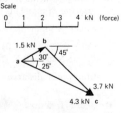

Fig 15

Problem 10 Use the parallelogram of forces method to find the magnitude and direction of the resultant of a force of 250 N acting at an angle of 135° and a force of 400 N acting at an angle of −120°.

From the procedure given in para. 6(c) and with reference to *Fig 16*:
(iv) **ab** is drawn at an angle of 135° and 250 units in length;
(v) **ac** is drawn at an angle of −120° and 400 units in length;
(vi) **bc** and **cd** are drawn to complete the parallelogram;
(vii) **ad** is drawn. By measurement **ad** is 413 units long at an angle of −156°.

That is, the resultant force is **413 N** at an angle of **−156°**.

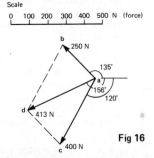

Fig 16

C. FURTHER PROBLEMS ON FORCES IN STATIC EQUILIBRIUM

(a) SHORT ANSWER PROBLEMS

1. Give one example of a scalar quantity and one example of a vector quantity.

2. Explain the difference between a scalar and a vector quantity.

3. What is meant by the centre of gravity of an object?

4. Where is the centre of gravity of a rectangular lamina?

5. What is meant by neutral equilibrium?

6. The moment of a force depends on two quantities. State these two quantities.

7. What is meant by the principle of moments?

8 State two conditions which must exist for an object to be in equilibrium?

9 State what is meant by a triangle of forces.

10 State what is meant by a parallelogram of forces.

(b) MULTI-CHOICE PROBLEMS (answers on page 148)

1 Which of the following statements is false?
   (a) Scalar quantities have size or magnitude only;
   (b) Vector quantities have both magnitude and direction;
   (c) Mass, length and time are all scalar quantities;
   (d) Distance, velocity and acceleration are all vector quantities.

2 If the centre of gravity of an object which is slightly disturbed is raised and the object returns to its original position when the disturbing force is removed, the object is said to be in
   (a) neutral equilibrium;   (c) static equilibrium;
   (b) stable equilibrium;   (d) unstable equilibrium.

3 Which of the following statements is false?
   (a) The centre of gravity of a lamina is at its point of balance;
   (b) The centre of gravity of a circular lamina is at its centre;
   (c) The centre of gravity of a rectangular lamina is at the point of intersection of its two sides;
   (d) The centre of gravity of a thin uniform rod is halfway along the rod.

4 A force of 10 N is applied at right angles to the handle of a spanner, 5 m from the centre of a nut. The moment on the nut is:
   (a) 50 N m; (b) 2 N/m, (c) 0.5 m/N, (d) 15 N m.

5 The distance d in *Fig 17* when the beam is in equilibrium is:
   (a) 0.5 m; (b) 1.0 m; (c) 4.0 m; (d) 15 m.

6 With reference to *Fig 18*, the clockwise moment about A is:
   (a) 70 N m, (b) 10 N m, (c) 60 N m; (d) $5 \times R_B$ N m.

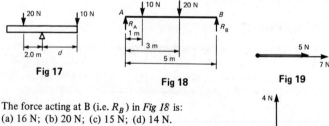

Fig 17

Fig 18

Fig 19

7 The force acting at B (i.e. R_B) in *Fig 18* is:
   (a) 16 N; (b) 20 N; (c) 15 N; (d) 14 N.

8 The force acting at A (i.e. R_A) in *Fig 18* is:
   (a) 16 N; (b) 10 N; (c) 15 N; (d) 14 N.

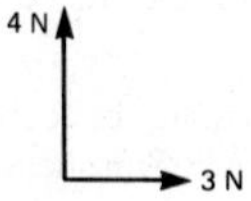

Fig 20

9 The magnitude of the resultant of the vectors shown in *Fig 19* is:
   (a) 2 N; (b) 12 N; (c) 35 N; (d) −2 N.

10 The magnitude of the resultant of the vectors shown in *Fig 20* is:
   (a) 7 N; (b) 5 N; (c) 1 N; (d) 12 N.

(c) CONVENTIONAL PROBLEMS

1. Determine the moment of a force of 25 N applied to a spanner at an effective length of 180 mm from the centre of a nut. [4.5 N m]

2. A moment of 7.5 N m is required to turn a wheel. If a force of 37.5 N is applied to the rim of the wheel, calculate the effective distance from the rim to the hub of the wheel. [200 mm]

3. Calculate the force required to produce a moment of 27 N m on a shaft, when the effective distance from the centre of the shaft to the point of application of the force is 180 mm. [150 N]

4. Determine distance d and the force acting at the support A for the force system shown in *Fig 21*, when the system is in equilibrium.
[50 mm; 3.8 kN]

5. If the 1 kN force shown in *Fig 21* is replaced by a force of F at a distance of 250 mm to the left of R_A, find the value of F for the system to be in equilibrium.
[560 N]

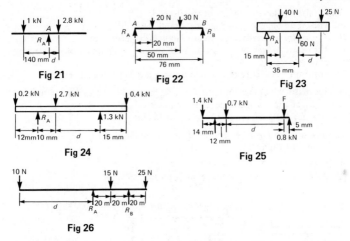

Fig 21　Fig 22　Fig 23
Fig 24　Fig 25
Fig 26

6. Determine the values of the forces acting at A and B for the force system shown in *Fig 22*. $[R_A = R_B = 25 \text{ N}]$

7. The forces acting on a beam are as shown in *Fig 23*. Neglecting the mass of the beam find the value of R_A and distance d when the beam is in equilibrium.
[5 N; 25 mm]

8. Calculate the force R_A and distance d for the beam shown in *Fig 24*. The mass of the beam should be neglected and equilibrium conditions assumed.
[2.0 kN; 24 mm]

9. For the force system shown in *Fig 25*, find the values of F and d for the system to be in equilibrium. [1.0 kN; 64 mm]

10. For the force system shown in *Fig 26*, determine distance d for forces R_A and R_B to be equal, assuming equilibrium conditions. [80 m]

In *Problems 11 to 20*, use a graphical method to determine the magnitude and direction of the resultant of the forces given.

11 1.3 kN and 2.7 kN, having the same line of action and acting in the same direction.
[4.0 kN in the direction of the forces]

12 470 N and 538 N having the same line of action but acting in opposite directions.
[68 N in the direction of the 538 N force]

13 13 N at 0° and 25 N at 30°. [36.8 N at 20°]

14 5 N at 60° and 8 N at 90°. [12.6 N at 79°]

15 1.3 kN at 45° and 2.8 kN at −30°. [3.4 kN at −8°]

16 1.7 N at 45° and 2.4 N at −60°. [2.6 kN at −20°]

17 9 N at 126° and 14 N at 223°. [15.7 N at −172°]

18 23.8 N at −50° and 14.4 N at 215°. [26.7 N at −82°]

19 0.7 kN at 147° and 1.3 kN at −71°. [0.86 kN at −100°]

20 47 N at 79° and 58 N at 247°. [15.5 N at −152°]

6 Pressure in fluids

A. MAIN POINTS CONCERNING PRESSURE IN FLUIDS

1 The **pressure** acting on a surface is defined as the perpendicular force per unit area of surface. The unit of pressure is the **pascal**, (Pa), where 1 pascal is equal to 1 newton per square metre. Thus

pressure, $p = \dfrac{F}{A}$ pascals

where F is the force in newtons acting at right angles to a surface of area A square metres.

When a force of 20 N acts uniformly over, and perpendicular to an area of 4 m², then the pressure on the area, p, is given by

$p = \dfrac{20 \text{ N}}{4 \text{ m}^2} = 5 \text{ Pa}$

2 A **fluid** is either a liquid or a gas and there are four basic factors governing the pressure within fluids.
  (a) The pressure at a given depth in a fluid is equal in all directions, see *Fig 1(a)*.
  (b) The pressure at a given depth in a fluid is independent of the shape of the container in which the fluid is held. In *Fig 1(b)*, the pressure at X is the same as the pressure as Y.

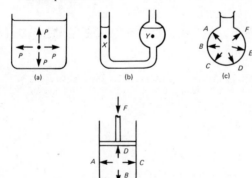

Fig 1

(c) Pressure acts at right angles to the surface containing the fluid. In *Fig 1(c)*, the pressures at points A to F all act at right angles to the container.

(d) When a pressure is applied to a fluid, this pressure is transmitted equally in all directions. In *Fig 1(d)*, if the mass of the fluid is neglected, the pressures at points A to D are all the same.

3 The pressure, p, at any point in a fluid depends on three factors:
(a) the density of the fluid, ρ, in kg/m³,
(b) the gravitational acceleration, g, taken as 9.8 m/s², and
(c) the height of fluid vertically above the point, h metres.

The relationship connecting these quantities is: $p = \rho g h$ pascals.

When the container shown in *Fig 2* is filled with water of density 1000 kg/m³, the pressure due to the water at a depth of 0.03 m below the surface is given by

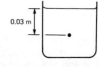

Fig 2

$p = \rho g h$
$= (1000 \times 9.8 \times 0.03)$ Pa
$= 294$ Pa.

4 The air above the earth's surface is a fluid, having a density, ρ, which varies from approximately 1.225 kg/m³ at sea level to zero in outer space. Since $p = \rho g h$, where height h is several thousands of metres, the air exerts a pressure on all points on the earth's surface. This pressure, called **atmospheric pressure**, has a value of approximately 100 kilopascals. Two terms are commonly used when measuring pressures:

(a) **absolute pressure**, meaning the pressure above that of an absolute vacuum, (i.e., zero pressure), and
(b) **gauge pressure**, meaning the pressure above that normally present due to the atmosphere. Thus:

absolute pressure = atmospheric pressure + gauge pressure

Thus, a gauge pressure of 50 kPa is equivalent to an absolute pressure of (100 + 50) kPa, i.e., 150 kPa, since the atmospheric pressure is approximately 100 kPa.

5 There are various ways of measuring pressure, and these include by:
(a) a U-tube manometer, (see *Problem 8*),
(b) a barometer, (see *Problem 9*),
(c) a pressure gauge, (see *Problem 10*).

B. WORKED PROBLEMS ON PRESSURE IN FLUIDS

Problem 1 A table loaded with books has a force of 250 N acting in each of its legs. If the contact area between each leg and the floor is 50 mm², find the pressure each leg exerts on the floor.

From para. 1, pressure $p = \dfrac{\text{force}}{\text{area}}$

Hence, $p = \dfrac{250 \text{ N}}{50 \text{ mm}^2} \times \dfrac{10^6 \text{ mm}^2}{1 \text{ m}^2}$

$= 5 \times 10^6 \text{ N/m}^2 = 5 \text{ MPa}$

That is, **the pressure exerted by each leg on the floor is 5 MPa.**

Problem 2 Calculate the force exerted by the atmosphere on a pool of water which is 30 m long by 10 m wide, when the atmospheric pressure is 100 kPa.

From para. 1, pressure = $\dfrac{\text{force}}{\text{area}}$

Hence, force = pressure × area

The area of the pool is 30 m × 10 m, i.e., 300 m².
Thus, force on pool, F = 100 kPa × 300 m²
and since 1 Pa = 1 N/m², $F = (100 \times 10^3) \dfrac{N}{m^2} \times 300 \text{ m}^2$

$$= 3 \times 10^7 \text{ N} = 30 \times 10^6 \text{ N} = 30 \text{ MN}.$$

That is, **the force on the pool of water is 30 MN.**

Problem 3 A circular piston exerts a pressure of 80 kPa on a fluid, when the force applied to the piston is 0.2 kN. Find the diameter of the piston.

From para. 1, pressure = $\dfrac{\text{force}}{\text{area}}$

Hence, area = $\dfrac{\text{force}}{\text{pressure}}$

Force in newtons is 0.2 kN × $\dfrac{1000 \text{ N}}{1 \text{ kN}}$ = 200 N.

Pressure in pascals is 80 kPa = 80 000 Pa = 80 000 N/m².

Hence, area = $\dfrac{200 \text{ N}}{80\,000 \text{ N/m}^2}$ = 0.0025 m².

Since the piston is circular, its area is given by $\dfrac{\pi d^2}{4}$

where d is the diameter of the piston.

Hence, area = $\dfrac{\pi d^2}{4}$ = 0.0025

$$d^2 = 0.0025 \times \dfrac{4}{\pi} = 0.003\,183$$

i.e. d = 0.0564 m, i.e., 56.4 mm.

Hence, the diameter of the piston is 56.4 mm.

Problem 4 A tank contains water to a depth of 600 mm. Calculate the water pressure (a) at a depth of 350 mm and (b) at the base of the tank. Take the density of water as 1000 kg/m³ and the gravitational acceleration as 9.8 m/s².

From para. 3, pressure p at any point in a fluid is given by $p = \rho g h$ pascals, where ρ is the density in kg/m³, g is the gravitational acceleration in m/s² and h is the height of fluid vertically above the point.
(a) At a depth of 350 mm, i.e. 0.35 m

p = 1000 × 9.8 × 0.35
   = 3430 Pa = **3.43 kPa.**

(b) At the base of the tank, the vertical height of the water is 600 mm, that is, 0.6 m. Hence

$p = 1000 \times 9.8 \times 0.6$
$= 5880$ Pa $= \mathbf{5.88}$ **kPa.**

Problem 5 A storage tank contains petrol to a height of 4.7 m. If the pressure at the base of the tank is 32.3 kPa, determine the density of the petrol. Take the gravitational acceleration as 9.8 m/s².

From para. 3, pressure $p = \rho g h$ pascals, where ρ is the density in kg/m³, g is the gravitational acceleration in m/s² and h is the vertical height of the petrol. Transposing gives

$\rho = \dfrac{p}{gh}$

The pressure p is 32.2 kPa, that is, 32 200 Pa.

Hence, density $\rho = \dfrac{32\ 200}{9.8 \times 4.7} = 699$ kg/m³.

That is, **the density of the petrol is 699 kg/m³**.

Problem 6 A vertical tube is partly filled with mercury of density 13 600 kg/m³. Find the height, in millimetres, of the column of mercury, when the pressure at the base of the tube is 101 kPa. Take the gravitational acceleration as 9.8 m/s².

From para. 3, pressure $p = \rho g h$, hence vertical height h is given by

$h = \dfrac{p}{\rho g}$

Pressure $p = 101$ kPa $= 101\ 000$ Pa.

Thus, $h = \dfrac{101\ 000}{13\ 600 \times 9.8} = 0.758$ m.

That is, **the height of the column of mercury is 758 mm.**

Problem 7 Calculate the absolute pressure at a point on a submarine, at a depth of 30 m below the surface of the sea, when the atmospheric pressure is 101 kPa. Take the density of sea water as 1030 kg/m³ and the gravitational acceleration as 9.8 m/s².

From para. 3, the pressure due to the sea, that is, the gauge pressure (p_g) is given by:
$p_g = \rho g h$ pascals
i.e. $p_g = 1030 \times 9.8 \times 30 = 302\ 820$ Pa
$= 302.82$ kPa.

From para. 4, absolute pressure = atmospheric pressure + gauge pressure
$= (101 + 302.82)$ kPa $= 403.82$ kPa

That is, **the absolute pressure at a depth of 30 m is 403.82 kPa.**

Problem 8 Briefly describe how pressure can be measured using a U-tube manometer and the effect of inclining one limb of the U-tube.

A manometer is a device used for measuring relatively small pressures, either above or below atmospheric pressure. A simple U-tube manometer is shown in *Fig 3*. Pressure p acting in, say, a gas main, pushes the liquid in the U-tube until equilibrium is obtained. At equilibrium: pressure in gas main, p = (atmospheric pressure, p_a) + (pressure due to the column of liquid, $\rho g h$) i.e. $p = p_a + \rho g h$

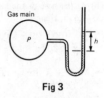

Fig 3

Thus, for example, if the atmospheric pressure, p_a, is 101 kPa, the liquid in the U-tube is water of density 1000 kg/m³ and height, h is 300 mm, then

absolute gas pressure = (101 000 + 1000 × 9.8 × 0.3) Pa
= (101 000 + 2940) Pa
= 103 940 Pa = 103.94 kPa

The gauge pressure of the gas is 2.94 kPa.

By filling the U-tube with a more dense liquid, say mercury having a density of 13 600 kg/m³, for a given height of U-tube, the pressure which can be measured is increased by a factor of 13.6.

By inclining one limb of the U-tube, as shown in *Fig 4*, greater sensitivity is achieved, that is, there is a larger movement of the liquid for a given change in pressure when compared with a U-tube having vertical limbs. An inclined manometer normally has a reservoir of sufficient area to give virtually a constant level

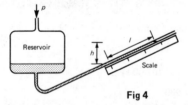

Fig 4

in the left-hand limb. From *Fig 4*, it can be seen that pressure p applied to the reservoir causes a scale change of 'l' in the inclined manometer compared with 'h' in a normal manometer.

Problem 9 Explain the principle of operation of a simple barometer and how atmospheric pressure can be measured using a Fortin barometer.

A simple barometer consists of a length of glass tubing, approximately 800 mm long and sealed at one end, which is filled with mercury and then inverted in a beaker of mercury, as shown in *Fig 5*. At equilibrium, the atmospheric pressure, p_a, is tending to force the mercury up the tube, whilst the force due to the column of mercury is tending to force the mercury out of the tube, i.e. $p_a = \rho g h$. As the atmospheric pressure varies, height h varies, giving an indication on the scale of the atmospheric pressure.

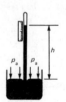

Fig 5

The **Fortin barometer** is as shown in *Fig 6*. Mercury is contained in a leather bag at the base of the mercury reservoir, and height, H, of the mercury in the reservoir can be adjusted using the screw at the base of the barometer to depress or release the leather bag. To measure the atmospheric pressure, the screw is adjusted until the pointer at H is just touching the surface of the mercury and the height of the mercury column is then read using the main and vernier scales. The measurement of atmospheric pressure using a Fortin barometer is achieved much more accurately than by using a simple barometer.

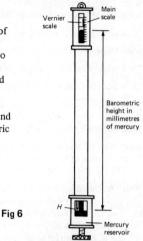

Fig 6

Problem 10 Explain briefly the operation of a Bourdon gauge for measuring pressure.

The main components of a **Bourdon pressure gauge** are shown in *Fig 7*. When pressure, p, is applied to the curved phosphor bronze tube, which is sealed at A, it tends to straighten, moving A to the right. Conversely, a decrease in pressure below that due to the atmosphere moves point A to the left. When A moves to the right, B moves to the left, rotating the pointer across a scale. This type of pressure gauge can be used to measure large pressures and pressures both above and below atmospheric pressure. The Bourdon pressure gauge indicates gauge pressure and is very widely used in industry for pressure measurements.

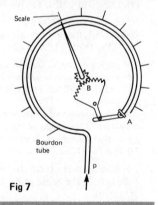

Fig 7

C. FURTHER PROBLEMS ON PRESSURE IN FLUIDS

(a) SHORT ANSWER PROBLEMS

1 Define pressure.

2 State the unit of pressure.

3 Define a fluid.

4 State the four basic factors governing the pressure in fluids.

5 Write down a formula for determining the pressure at any point in a fluid in symbols, defining each of the symbols and giving their units.

6 What is meant by atmospheric pressure?

7 State the approximate value of atmospheric pressure.

8 State what is meant by gauge pressure.

9 State what is meant by absolute pressure.

10 State the relationship between absolute, gauge and atmospheric pressures.

(b) MULTI-CHOICE PROBLEMS (answers on page 148)

1 A force of 50 N acts uniformly over and at right angles to a surface. When the area of the surface is 5 m^2, the pressure on the area is:
(a) 250 Pa; (b) 10 Pa; (c) 45 Pa; (d) 55 Pa.

2 Which of the following statements is false?
The pressure at a given depth in a fluid
(a) is equal in all directions;
(b) is independent of the shape of the container;
(c) acts at right angles to the surface containing the fluid;
(d) depends on the area of the surface.

3 A container holds water of density 1000 kg/m^3. Taking the gravitational acceleration as 10 m/s^2, the pressure at a depth of 100 mm is:
(a) 1 kPa; (b) 1 MPa; (c) 100 Pa; (d) 1 Pa.

4 If the water in *Problem 3* is now replaced by a fluid having a density of 2000 kg/m^3, the pressure at a depth of 100 mm is:
(a) 2 kPa; (b) 500 kPa; (c) 200 Pa; (d) 0.5 Pa.

5 The gauge pressure of fluid in a pipe is 70 kPa and the atmospheric pressure is 100 kPa. The absolute pressure of the fluid in the pipe is
(a) 7 MPa; (b) 30 kPa; (c) 170 kPa; (d) $\frac{10}{7}$ kPa.

6 A U-tube manometer contains mercury of density 13 600 kg/m^3. When the difference in the height of the mercury levels is 100 mm and taking the gravitational acceleration as 10 m/s^2, the gauge pressure is
(a) 13.6 Pa; (b) 13.6 MPa; (c) 13 710 Pa; (d) 13.6 kPa

7 The mercury in the U-tube of *Problem 6* is to be replaced by water of density 1000 kg/m^3. The height of the tube to contain the water for the same gauge pressure is:

(a) $\frac{1}{13.6}$ of the original height;

(b) 13.6 times the original height;

(c) 13.6 m more than the original height;

(d) 13.6 m less than the original height.

(c) CONVENTIONAL PROBLEMS

Take the gravitational acceleration as 9.8 m/s^2, the density of water as 1000 kg/m^3 and the density of mercury as 13 600 kg/m^3.

1 A force of 280 N is applied to a piston of a hydraulic system of cross-sectional

area 0.010 m². Determine the pressure produced by the piston in the hydraulic fluid. [28 kPa]

2 Find the force on the piston of *Problem 1* to produce a pressure of 450 kPa. [4.5 kN]

3 If the area of the piston in *Problem 1* is halved and the force applied is 280 N, determine the new pressure in the hydraulic fluid. [56 kPa]

4 Determine the pressure acting at the base of a dam, when the surface of the water is 35 m above base level. [343 kPa]

5 An uncorked bottle is full of sea water of density 1030 kg/m³.
Calculate, correct to 3 significant figures, the pressures on the side wall of the bottle at depths of (a) 30 mm and (b) 70 mm below the top of the bottle.
[(a) 303 Pa; (b) 707 Pa]

6 A U-tube manometer is used to determine the pressure at a depth of 500 mm below the free surface of a fluid. If the pressure at this depth is 6.86 kPa, calculate the density of the liquid used in the manometer. [1400 kg/m³]

7 The height of a column of mercury in a barometer is 750 mm. Determine the atmospheric pressure, correct to 3 significant figures. [100 kPa]

8 A U-tube manometer containing mercury gives a height reading of 250 mm of mercury when connected to a gas cylinder. If the barometer reading at the same time is 756 mm of mercury, calculate the absolute pressure of the gas in the cylinder, correct to 3 significant figures. [134 kPa]

9 A water manometer connected to a condenser shows that the pressure in the condenser is 350 mm below atmospheric pressure. If the barometer is reading 760 mm of mercury, determine the absolute pressure in the condenser, correct to 3 significant figures. [97.9 kPa]

10 A Bourdon pressure gauge shows a pressure of 1.151 MPa. If the absolute pressure is 1.25 MPa, find the atmospheric pressure in millimetres of mercury. [743 mm]

11 Explain why barometers are usually filled with mercury rather than water.

12 Explain how atmospheric pressure may be measured, using a Fortin barometer.

7 Speed and velocity

A. MAIN POINTS CONCERNING SPEED AND VELOCITY

1 Speed is the rate of covering distance and is given by:

$$\text{speed} = \frac{\text{distance travelled}}{\text{time taken}}$$

The usual units for speed are metres per second, (m/s or m s^{-1}), or kilometres per hour, (km/h or km h^{-1}). Thus if a person walks 5 kilometres in 1 hour, the speed of the person is $\frac{5}{1}$, that is, 5 kilometres per hour.

The symbol for the SI unit of speed and velocity is written as m s^{-1}, called the 'index notation'. However, engineers usually use the symbol m/s, called the 'oblique notation', and it is this notation which is largely used in this chapter and other chapters on mechanics. One of the exceptions is when labelling the axes of graphs, when two obliques occur, and in this case the index notation is used. Thus for speed or velocity, the axis markings are speed/m s^{-1} or velocity/m s^{-1}.

2 One way of giving data on the motion of an object is graphically. A graph of distance travelled, (the scale on the vertical axis of the graph), against time, (the scale on the horizontal axis of the graph), is called a **distance-time graph**. Thus if a plane travels 500 kilometres in its first hour of flight and 750 kilometres in its second hour of flight, then after 2 hours, the total distance travelled is (500+750) kilometres, that is, 1250 kilometres. The distance-time graph for this flight is shown in *Fig 1*.

3 The **average speed** is given by

$$\frac{\text{total distance travelled}}{\text{total time taken}}$$

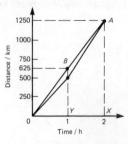

Fig 1

Thus, the average speed of the plane in para. 2 is:

$$\frac{(500 + 750) \text{ km}}{(1 + 1) \text{ h}}, \text{ i.e. } \frac{1250}{2} \text{ or } 625 \text{ km/h}.$$

If points 0 and A are joined in *Fig 1*, the slope of line 0A is defined as

change in distance (vertical)
change in time (horizontal)

for any two points on line 0A. For point A, the change in distance is AX, that is 1250 kilometres, and the change in time is 0X, that is, 2 hours. Hence the average speed is

$\frac{1250}{2}$, i.e. 625 kilometres per hour.

Alternatively, for point B on line 0A, the change in distance is BY, that is, 625 kilometres and the change in time is 0Y, that is 1 hour, hence the average speed is 625/1, i.e. 625 kilometres per hour.

In general, the average speed of an object travelling between points M and N is given by the slope of line MN on the distance-time graph.

4 The **velocity** of an object is the speed of the object **in a specified direction**. Thus, if a plane is flying due south at 500 kilometres per hour, its speed is 500 kilometres per hour, but its velocity is 500 kilometres per hour **due south**. It follows that if the plane had flown in a circular path for one hour at a speed of 500 kilometres per hour, so that one hour after taking off it is again over the airport, its average velocity in the first hour of flight is zero.

5 The **average velocity** is given by:

distance travelled in a specific direction
time taken

If a plane flies from place O to place A, a distance of 300 kilometres in one hour, A being due north of O, then OA in *Fig 2* represents the first hour of flight. It then flies from A to B, a distance of 400 kilometres during the second hour of flight, B being due east of A, thus AB in *Fig 2* represents its second hour of flight. Its average velocity for the two hour flight is

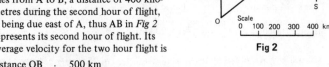

Fig 2

$\frac{\text{distance OB}}{2 \text{ hours}}$ i.e. $\frac{500 \text{ km}}{2 \text{ h}}$

or 250 km/h in direction OB.

6 A graph of velocity, (scale on the vertical axis), against time, (scale on the horizontal axis), is called a **velocity-time graph**. The graph shown in *Fig 3* represents a plane flying for 3 hours at a constant speed of 600 kilometres per hour in a specified direction. The shaded area represents velocity, (vertically), multiplied by time (horizontally), and has units of

$\frac{\text{kilometres}}{\text{hours}} \times \text{hours}$,

i.e. kilometres, and represents the distance travelled in a specific direction. Another method of determining the distance travelled is from:

distance travelled = average velocity × time.

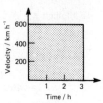

Fig 3

Thus if a plane travels due south at 600 kilometres per hour for 20 minutes, the distance covered is

$$\frac{600 \text{ km}}{1 \text{ h}} \times \frac{20}{60} \text{ h, i.e. } 200 \text{ km.}$$

B. WORKED PROBLEMS ON SPEED AND VELOCITY

Problem 1 A man walks 600 metres in 5 minutes. Determine his speed in (a) metres per second and (b) kilometres per hour.

(a) Speed = $\frac{\text{distance travelled}}{\text{time taken}} = \frac{600 \text{ m}}{5 \text{ min}}$

$= \frac{600 \text{ m}}{5 \text{ min}} \times \frac{1 \text{ min}}{60 \text{ s}} = \textbf{2 m/s}$

(b) $\frac{2 \text{ m}}{1 \text{ s}} = \frac{2 \text{ m}}{1 \text{ s}} \times \frac{1 \text{ km}}{1000 \text{ m}} \times \frac{3600 \text{ s}}{1 \text{ h}} = \textbf{7.2 km/h.}$

(*Note*: to change from m/s to km/h, multiply by 3.6)

Problem 2 A car travels at 50 kilometres per hour for 24 minutes. Find the distance travelled in this time.

Since speed = $\frac{\text{distance travelled}}{\text{time taken}}$

Then,

distance travelled = speed × time taken.

Time = 24 minutes = $\frac{24}{60}$ hours, hence

distance travelled = $50 \frac{\text{km}}{\text{h}} \times \frac{24}{60} \text{h} = \textbf{20 km.}$

Problem 3 A train is travelling at a constant speed of 25 metres per second for 16 kilometres. Find the time taken to cover this distance.

Since speed = $\frac{\text{distance travelled}}{\text{time taken}}$

Then,

time taken = $\frac{\text{distance travelled}}{\text{speed}}$

16 kilometres = 16 000 metres

Hence,

time taken = $\frac{16\ 000 \text{ m}}{\frac{25 \text{ m}}{1 \text{ s}}} = 16\ 000 \text{ m} \times \frac{1 \text{ s}}{25 \text{ m}}$

$= 640 \text{ s}$

$640 \text{ s} = 640 \text{ s} \times \frac{1 \text{ min}}{60 \text{ s}} = \textbf{10}\frac{\textbf{2}}{\textbf{3}}\textbf{min}$

Problem 4 A person travels from point O to A, then from A to B and finally from B to C. The distances of A, B and C from O and the times, measured from the start to reach points A, B and C are as shown.

	A	B	C
Distance (m)	100	200	250
Time (s)	40	60	100

Plot the distance-time graph and determine the speed of travel for each of the three parts of the journey.

The vertical scale of the graph is distance travelled and the scale is selected to span 0 to 250 m, the total distance travelled from the start. The horizontal scale is time and spans 0 to 100 seconds, the total time taken to cover the whole journey. Co-ordinates corresponding to A, B and C are plotted and OA, AB and BC are joined by straight lines. The resulting distance-time graph is shown in *Fig 4*.

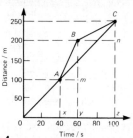

Fig 4

The speed is given by the slope of the distance-time graph.

Speed for part OA of the journey = slope of OA = $\frac{Ax}{Ox}$

$$= \frac{100 \text{ m}}{40 \text{ s}} = 2\frac{1}{2} \text{ m/s}$$

Speed for part AB of the journey = slope of AB = $\frac{Bm}{Am}$

$$= \frac{(200 - 100) \text{ m}}{(60 - 40) \text{ s}} = \frac{100 \text{ m}}{20 \text{ s}}$$

$$= 5 \text{ m/s}.$$

Speed for part BC of the journey = slope of BC = $\frac{Cn}{Bn}$

$$= \frac{(250 - 200) \text{ m}}{(100 - 60) \text{ s}}$$

$$= \frac{50 \text{ m}}{40 \text{ s}} = 1\frac{1}{4} \text{ m/s}.$$

Problem 5 Determine the average speed (both in m/s and km/h) for the whole journey for the information given in *Problem 4*.

Average speed = $\frac{\text{total distance travelled}}{\text{total time taken}}$ = slope of line OC.

From *Fig 4*, slope of line OC = $\frac{Cz}{Oz} = \frac{250 \text{ m}}{100 \text{ s}} = 2.5$ m/s

$$2.5 \text{ m/s} = \frac{2.5 \text{ m}}{1 \text{ s}} \times \frac{1 \text{ km}}{1000 \text{ m}} \times \frac{3600 \text{ s}}{1 \text{ h}}$$

$$= 2.5 \times 3.6 \text{ km/h} = 9 \text{ km/h}.$$

Problem 6 A coach travels from town A to town B, a distance of 40 kilometres at an average speed of 55 kilometres per hour. It then travels from town B to town C, a distance of 25 kilometres in 35 minutes. Finally, it travels from town C to town D at an average speed of 60 kilometres per hour in 45 minutes. Determine:
(a) the time taken to travel from A to B;
(b) the average speed of the coach from B to C;
(c) the distance from C to D; and
(d) the average speed of the whole journey from A to D.

(a) From town A to town B:

Since speed = $\dfrac{\text{distance travelled}}{\text{time taken}}$

then,

time taken = $\dfrac{\text{distance travelled}}{\text{speed}}$

$= \dfrac{40 \text{ km}}{\dfrac{55 \text{ km}}{1 \text{ h}}} = 40 \text{ km} \times \dfrac{1 \text{ h}}{55 \text{ km}}$

$= \dfrac{8}{11} \text{ h} \approx$ **43.6 min.**

(b) From town B to town C:

Since speed = $\dfrac{\text{distance travelled}}{\text{time taken}}$ and 35 min = $\dfrac{35}{60}$ h

Then,

speed = $\dfrac{25 \text{ km}}{\dfrac{35}{60} \text{ h}} = \dfrac{25 \times 60}{35}$ km/h

$\approx$ **42.86 km/h.**

(c) From town C to town D:

Since speed = $\dfrac{\text{distance travelled}}{\text{time taken}}$,

Then,
distance travelled = speed × time taken.

45 min = $\dfrac{3}{4}$ h, hence,

distance travelled = $60 \dfrac{\text{km}}{\text{h}} \times \dfrac{3}{4} \text{h} =$ **45 km.**

(d) From town A to town D:

Average speed = $\dfrac{\text{total distance travelled}}{\text{total time taken}}$

$= \dfrac{(40 + 25 + 45) \text{ km}}{\left(\dfrac{43.6}{60} + \dfrac{35}{60} + \dfrac{45}{60}\right) \text{ h}}$

$= \dfrac{110 \text{ km}}{\dfrac{123.6}{60} \text{ h}} = \dfrac{110 \times 60}{123.6}$ km/h

$\approx$ **53.4 km/h.**

Problem 7 The motion of an object is described by the speed-time graph given in *Fig 5*. Determine the distance covered by the object when moving from O to B.

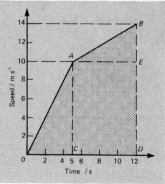

Fig 5

The distance travelled is given by the area beneath the speed-time graph, shown shaded in *Fig 5*.

Area of triangle OAC $= \frac{1}{2} \times$ base $\times$ perpendicular height

$$= \frac{1}{2} \times 5 \text{ s} \times 10 \, \frac{\text{m}}{\text{s}} = 25 \text{ m}.$$

Area of rectangle AEDC = base $\times$ height

$$= (12 - 5) \text{ s} \times (10 - 0) \, \frac{\text{m}}{\text{s}} = 70 \text{ m}.$$

Area of triangle ABE $= \frac{1}{2} \times$ base $\times$ perpendicular height

$$= \frac{1}{2} \times (12 - 5) \text{ s} \times (14 - 10) \, \frac{\text{m}}{\text{s}}$$

$$= \frac{1}{2} \times 7 \text{ s} \times 4 \, \frac{\text{m}}{\text{s}} = 14 \text{ m}.$$

Hence, distance covered by the object moving from O to B is

$(25 + 70 + 14)$ m, i.e. **109 m**.

C. FURTHER PROBLEMS ON SPEED AND VELOCITY

(a) SHORT ANSWER PROBLEMS

1 Speed is defined as . . . .

2 Speed is given by $\frac{\cdots\cdots\cdots\cdots\cdots}{\cdots\cdots\cdots\cdots\cdots}$.

3 The usual units for speed are . . . . or . . . .

4 Average speed is given by $\frac{\cdots\cdots\cdots\cdots\cdots}{\cdots\cdots\cdots\cdots\cdots}$.

5 The velocity of an object is . . . .

6 Average velocity is given by $\frac{\cdots\cdots\cdots\cdots\cdots}{\cdots\cdots\cdots\cdots\cdots}$.

7 The area beneath a velocity-time graph represents the . . . .

8 Distance travelled = ... × ....

9 The slope of a distance-time graph gives the ....

10 The average speed can be determined from a distance-time graph from
   ....

(b) MULTI-CHOICE PROBLEMS (answers on page 148)

An object travels for 3 s at an average speed of 10 m/s and then for 5 s at an average speed of 15 m/s. In *Problems 1 to 3*, select the correct answers from those given below.

(a) 105 m/s; (b) 3 m; (c) 30 m; (d) $13\frac{1}{8}$ m/s; (e) $3\frac{1}{3}$ m; (f) $\frac{3}{10}$ m;

(g) 75 m; (h) $\frac{1}{3}$ m; (i) $12\frac{1}{2}$ m/s.

1 The distance travelled in the first 3 s.

2 The distance travelled in the latter 5 s.

3 The average speed over the 8 s period.

4 Which of the following statements is false?
   (a) Speed is the rate of covering distance;
   (b) Speed and velocity are both measured in m/s units;
   (c) Speed is the velocity of travel in a specified direction;
   (d) The area beneath the velocity-time graph gives distance travelled;

In *Problems 5 to 7*, use the table to obtain the quantities stated, selecting the correct answer from those given below.

Distance	Time	Speed
20 m	30 s	X
5 km	Y	20 km/h
Z	3 min	10 m/min

(a) 30 m; (b) $\frac{1}{4}$ h; (c) 600 m/s; (d) $\frac{10}{3}$ m; (e) $\frac{2}{3}$ m/s; (f) $\frac{3}{10}$ m; (g) 4 h;

(h) $1\frac{1}{4}$ m/s; (i) 100 h.

5 Quantity X.

6 Quantity Y.

7 Quantity Z.

In *Problems 8 to 10*, refer to the distance-time graph shown in *Fig 6*. Select the correct answer from those given below.

(a) $\frac{3}{2}$ m/s; (b) 3 m/s; (c) $\frac{1}{2}$ m/s;

(d) 2 m/s; (e) $\frac{4}{5}$ m/s; (f) 6 m/s;

(g) $\frac{5}{4}$ m/s; (h) $\frac{2}{3}$ m/s; (i) $\frac{1}{3}$ m/s.

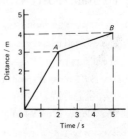

Fig 6

8 The average speed when travelling from O to A.

9 The average speed when travelling from A to B.

10 The average overall speed when travelling from O to B.

(c) CONVENTIONAL PROBLEMS

1 A train covers a distance of 96 km in $1\frac{1}{3}$ h. Determine the average speed of the
   train (a) in km/h and (b) in m/s. [(a) 72 km/h; (b) 20 m/s]

2 A horse trots at an average speed of 12 km/h for 18 minutes; determine the
   distance covered by the horse in this time. [3.6 km]

3 A ship covers a distance of 1365 km at an average speed of 15 km/h. How long
   does it take to cover this distance? [3 days 19 hours]

4 Using the information given in the distance-time graph shown in *Fig 7*, determine
   the average speed when travelling from O to A, A to B, B to C, O to C and A to C.
   [O to A, 30 km/h; A to B, 40 km/h; B to C, 10 km/h; O to C, 24 km/h; A to C, 20 km/h]

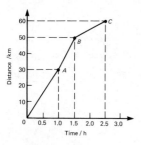
Fig 7

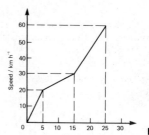
Fig 8

5 The distances travelled by an object from point O and the corresponding times
   taken to reach A, B, C and D respectively from the start are as shown.

Points	Start	A	B	C	D
Distance (m)	0	20	40	60	80
Time (s)	0	5	12	18	25

   Draw the distance-time graph and hence determine the average speeds from
   O to A, A to B, B to C, C to D and O to D.

   $[4 \text{ m/s}; 2\frac{6}{7} \text{ m/s}; 3\frac{1}{3} \text{ m/s}; 2\frac{6}{7} \text{ m/s}; 3\frac{1}{5} \text{ m/s}]$

6 A train leaves station A and travels via stations B and C to station D. The times
   the train passes the various stations are as shown.

Station	A	B	C	D
Times	10.55 am	11.40 am	12.15 pm	12.50 pm

   The average speeds are:

   A to B, 56 km/h;
   B to C, 72 km/h; and
   C to D, 60 km/h.

   Calculate the total distance from A to D. [119 km]

7   A gun is fired 5 km north of an observer and the sound takes 15 s to reach him. Determine the average velocity of sound waves in air at this place.

$[333\frac{1}{3}$ m/s or 1200 km/h]

8   The light from a star takes 2½ years to reach an observer. If the velocity of light is 330×10^6 m/s, determine the distance of the star from the observer in kilometres, based on a 365 day year. $[2.6 \times 10^{13}$ km]

9   The speed-time graph for a car journey is shown in *Fig 8*. Determine the distance travelled by the car.

$[12\frac{1}{2}$ km]

10 The motion of an object is as follows:

A to B, distance 122 m, time 64 s;
B to C, distance 80 m at an average speed of 20 m/s;
C to D, time 7 s at an average speed of 14 m/s.

Determine the overall average speed of the object when travelling from A to D.

[4 m/s]

8 Acceleration and force

A. MAIN POINTS CONCERNING ACCELERATION AND FORCE

1 **Acceleration** is the rate of change of speed or velocity with time. The average acceleration, a, is given by:

$$a = \frac{\text{change in velocity}}{\text{time taken}}.$$

The usual units are metres per second squared, (m/s² or m s⁻²). If u is the initial velocity of an object in metres per second, v is the final velocity in metres per second and t is the time in seconds elapsing between the velocities of u and v, then

average acceleration, $a = \dfrac{v - u}{t}$ m/s².

2 A graph of speed, (scale on the vertical axis), against time, (scale on the horizontal axis) is called a **speed-time graph**. For the speed-time graph shown in *Fig 1*, the slope of line OA is given by AX/OX. AX is the change in velocity from an initial velocity u of zero to a final velocity, v, of 4 metres per second. OX is the time taken for this change in velocity, thus

$$\frac{\text{AX}}{\text{OX}} = \frac{\text{change in velocity}}{\text{time taken}} = \text{the acceleration in the first two seconds.}$$

From the graph:
$$\frac{\text{AX}}{\text{OX}} = \frac{4 \text{ m/s}}{2 \text{ s}} = 2 \text{ m/s}^2$$

i.e. the acceleration is 2 m/s².
Similarly, the slope of line AB in *Fig 1* is given by BY/AY, i.e. the acceleration between 2 and 5 s is
$$\frac{8 - 4}{5 - 2} = \frac{4}{3} = 1\frac{1}{3} \text{ m/s}^2$$

In general, the slope of a line on a speed-time graph gives the acceleration.

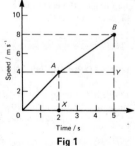

Fig 1

3 If a dense object such as a stone is dropped from a height, called **free fall**, it has a constant acceleration of approximately 9.8 m/s². In a vacuum, all objects have this same constant acceleration vertically downwards, that is, a feather has the

same acceleration as a stone. However, if free fall takes place in air, dense objects have the approximately constant acceleration of 9.8 m/s² over short distances, but objects which have a low density, such as feathers, have little or no acceleration.

4 For bodies moving with a constant acceleration, the average acceleration is the constant value of the acceleration, and since from para. 1,

$a = \frac{v - u}{t}$, then $a \times t = v - u$ or $\boldsymbol{v = u + at}$,

where
u is the initial velocity in m/s,
v is the final velocity in m/s,
a is the constant acceleration in m/s²,
t is the time in s.

When symbol 'a' has a negative value, it is called **deceleration** or **retardation**. The equation, $v = u + at$ is called an **equation of motion**.

5 When an object is pushed or pulled, a **force** is applied to the object. This force is measured in **newtons**, (N). The effects of pushing or pulling an object are:
  (i) to cause a change in the motion of the object, and
  (ii) to cause a change in the shape of the object.

If a change occurs in the motion of the object, that is, its speed changes from u to v, then the object accelerates. Thus, it follows that acceleration results from a force being applied to an object. If a force is applied to an object and it does not move, then the object changes shape, that is, deformation of the object takes place. Usually the change in shape is so small that it cannot be detected by just watching the object. However, when very sensitive measuring instruments are used, very small changes in dimensions can be detected.

6 A force of attraction exists between all objects. The factors governing the size of this force are the masses of the objects and the distances between their centres

$F \propto \frac{m_1 m_2}{d^2}$

Thus, if a person is taken as one object and the earth as a second object, a force of attraction exists between the person and the earth. This force is called the **gravitational force** and is the force which gives a person a certain weight when standing on the earth's surface. It is also this force which gives freely falling objects a constant acceleration in the absence of other forces.

B. WORKED PROBLEMS ON ACCELERATION AND FORCE

Problem 1 The speed of a car travelling along a straight road changes uniformly from zero to 50 km/h in 20 s. It then maintains this speed for 30 s and finally reduces speed uniformly to rest in 10 s. Draw the speed-time graph for this journey.

The vertical scale of the speed-time graph is speed, (km h⁻¹) and the horizontal scale is time, (s). Since the car is initially at rest, then at time 0 seconds, the speed is 0 km/h. After 20 s, the speed if 50 km/h, which corresponds to point A on the speed-time graph shown in *Fig 2*. Since the change in speed is uniform, a straight line is drawn joining points O and A. The speed is constant at 50 km/h for the next

30 s, hence, horizontal line AB is drawn in
Fig 2 for the time period 20 s to 50 s.

Finally, the speed falls from 50 km/h
at 50 s to zero in 10 s, hence point C on the
speed-time graph in *Fig 2* corresponds to a
speed of zero and a time of 60 s. Since the
reduction in speed is uniform, a straight
line is drawn joining BC. Thus, the speed-
time graph for the journey is as shown in
Fig 2.

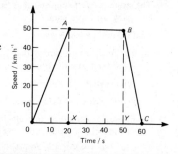

Fig 2

Problem 2 For the speed-time graph shown in *Fig 2*, find the accelerations for each of the three stages of the journey.

From para. 2, the slope of line OA gives the uniform acceleration for the first 20 s of the journey.

Slope of OA = $\dfrac{AX}{OX} = \dfrac{(50 - 0)\text{km/h}}{(20 - 0)\text{ s}} = 50 \dfrac{\text{km}}{\text{h}} \times \dfrac{1}{20 \text{ s}}$

Expressing 50 km/h in metre-second units, gives:

$50 \dfrac{\text{km}}{\text{h}} = \dfrac{50 \text{ km}}{1 \text{ h}} \times \dfrac{1000 \text{ m}}{1 \text{ km}} \times \dfrac{1 \text{ h}}{3600 \text{ s}}$

$= \dfrac{50}{3.6} \text{ m/s}$

[*Note*: to change from km/h to m/s, divide by 3.6]

Thus, 50 km/h $\times \dfrac{1}{20 \text{ s}} = \dfrac{50}{3.6} \text{ m/s} \times \dfrac{1}{20 \text{ s}}$

$= 0.694 \text{ m/s}^2$,

i.e. the acceleration during the first 20 s is 0.694 m/s².

Acceleration is defined as $\dfrac{\text{change of velocity}}{\text{time taken}}$, or as $\dfrac{\text{change of speed}}{\text{time taken}}$,

Since the car is travelling along a straight road. Since there is no change in speed for the next 30 s, (line AB in *Fig 2* is horizontal), **then the acceleration for this period is zero.**

From para. 2, the slope of line BC gives the uniform deceleration for the final 10 s of the journey.

Slope of BC = $\dfrac{BY}{YC} = \dfrac{50 \text{ km/h}}{10 \text{ s}} = \dfrac{50 \text{ m}}{3.6 \text{ s}} \times \dfrac{1}{10 \text{ s}}$

$= 1.39 \text{ m/s}^2$,

i.e. the deceleration during the final 10 s is 1.39 m/s².
Alternatively, **the acceleration is -1.39 m/s².**

Problem 3 A stone is dropped from a plane. Determine (a) its velocity after 2 s and (b) the increase in velocity during the third second, in the absence of all forces except that due to gravity.

The stone is free falling and thus has an acceleration, a, of approximately 9.8 m/s^2, (see para. 3). From para. 4,

final velocity, $v = u + at$.

(a) The initial downward velocity of the stone, u, is zero. The acceleration, a, is 9.8 m/s^2 and the time during which the stone is accelerating is 2 s. Hence,

final velocity, $v = 0 + 9.8 \times 2 = 19.6$ m/s.

i.e. **the velocity of the stone after 2 s is approximately 19.6 m/s.**

(b) From part (a) above, the velocity after two seconds, u, is 19.6 m/s. The velocity after 3 s, applying $v = u + at$, is

$v \approx 19.6 + 9.8 \times 3 \approx 49$ m/s.

Thus, **the change in velocity during the third second is**

$(49 - 19.6)$ m/s, that is, approximately **29.4 m/s.**

(Since the value, $a = 9.8$ m/s^2 is only an approximate value, then the answer can only be an approximate value.)

Problem 4 Determine how long it takes an object, which is free falling, to change its speed from 100 km/h to 150 km/h, assuming all other forces, except that due to gravity, are neglected.

The initial velocity, u, is 100 km/h, i.e. $\dfrac{100}{3.6}$ m/s, (see *Problem 2*).

The final velocity, v, is 150 km/h, i.e. $\dfrac{150}{3.6}$ m/s. Since the object is free falling, the acceleration, a, is approximately 9.8 m/s^2. From para. 4,

$v = u + at$

i.e. $\dfrac{150}{3.6} = \dfrac{100}{3.6} + 9.8 \times t$

Transposing, gives $9.8 \times t = \dfrac{150 - 100}{3.6} = \dfrac{50}{3.6}$

Hence, $t = \dfrac{50}{3.6 \times 9.8} \approx 1.42$ s.

Since the value of a is only approximate, and rounding-off errors have occurred in calculations, then **the approximate time for the velocity to change from 100 km/h to 150 km/h is 1.42 s.**

Problem 5 A train travelling at 30 km/h accelerates uniformly to 50 km/h in 2 minutes. Determine the acceleration.

---

30 km/h = $\dfrac{30}{3.6}$ m/s, (see *Problem 2*)

50 km/h = $\dfrac{50}{3.6}$ m/s

2 min = $2 \times 60 = 120$ s.

From para. 4, $v = u + at$,

i.e. $\dfrac{50}{3.6} = \dfrac{30}{3.6} + a \times 120$

Transposing, gives $120 \times a = \dfrac{50-30}{3.6}$

$$a = \dfrac{20}{3.6 \times 120} = 0.0463 \text{ m/s}^2$$

i.e. **the uniform acceleration of the train is 0.0463 m/s².**

Problem 6 A car travelling at 50 km/h applies its brakes for 6 s and decelerates uniformly at 0.5 m/s. Determine its speed in km/h after the 6 s braking period.

The initial velocity, $u = 50$ km/h $= \dfrac{50}{3.6}$ m/s, (see *Problem 2*).

From para. 4, $v = u + at$.

Since the car is decelerating, i.e., it has a negative acceleration, then $a = -0.5$ m/s² and t is 6 s.

Thus, final velocity, $v = \dfrac{50}{3.6} + (-0.5)(6)$

$= 13.\dot{8} - 3 = 10.\dot{8}$ m/s

$10.\dot{8}$ m/s $= 10.\dot{8} \dfrac{\text{m}}{\text{s}} \times \dfrac{1 \text{ km}}{1000 \text{ m}} \times \dfrac{3600 \text{ s}}{1 \text{ h}}$

$= 10.\dot{8} \times 3.6$ km/h $= 39.2$ km/h.

[*Note*: to convert m/s to km/h, multiply by 3.6]

Thus, the speed after braking is 39.2 km/h.

Problem 7 A cyclist accelerates uniformly at 0.3 m/s² for 10 s, and his speed after accelerating is 20 km/h. Find his initial speed.

The final speed, v is $\dfrac{20}{3.6}$ m/s.

Time, t is 10 s.
Acceleration, a is 0.3 m/s².
From para. 4, $v = u + at$, where u is the initial speed.

Hence, $\dfrac{20}{3.6} = u + 0.3 \times 10$

$u = \dfrac{20}{3.6} - 3 = 2.\dot{5}$ m/s

$2.\dot{5}$ m/s $= 2.\dot{5} \times 3.6$ km/h, (see *Problem 6*) $= 9.2$ km/h.

i.e. **the initial speed of the cyclist is 9.2 km/h.**

C. FURTHER PROBLEMS ON ACCELERATION AND FORCE

(a) SHORT ANSWER PROBLEMS

1   Acceleration is defined as . . . .

2 Acceleration is given by $\frac{\text{................}}{\text{................}}$.

3 The usual units for acceleration are ....

4 The slope of a speed-time graph gives the ....

5 The value of free-fall acceleration for a dense object is approximately ....

6 The relationship between initial velocity, u, final velocity v, acceleration, a, and time, t, is ....

7 A negative acceleration is called a .... or a ....

8 Force is measured in ....

9 The two effects of pushing or pulling an object are .... or ....

10 A gravitational force gives free falling objects a .... , in the absence of all other forces.

(b) MULTI-CHOICE PROBLEMS (answers on page 148)

Ten statements, (a) to (j), are given below, some of the statements being true and the remainder false.

(a) Acceleration is the rate of change of speed or velocity with distance.

(b) Average acceleration = $\frac{\text{change of velocity}}{\text{time taken}}$.

(c) Average acceleration = $\frac{u - v}{t}$, where u is the initial velocity, v is the final velocity and t is the time.

(d) The slope of a speed-time graph gives the acceleration.

(e) The acceleration of a dense object during free fall is approximately 9.8 m/s^2 in the absence of all other forces except gravity.

(f) When the initial and final velocities are u and v respectively, a is the acceleration and t the time, then $u = v + at$.

(g) Force is measured in newtons.

(h) When an object is pulled or pushed there is either a change in shape or a change in force.

(i) The force of attraction between an object on the earth's surface and the earth is called the gravitational force.

(j) Gravitational force gives free falling objects an increasing acceleration in the absence of all other forces.

In *Problems 1 to 5*, select the statements required from those given.

1 (b), (g), (i), (j). Which statement is false?

2 (a), (d), (e), (g). Which statement is false?

3 (b), (c), (f), (h). Which statement is true?

4 (a), (c), (d), (j). Which statement is true?

5 (b), (e), (f), (i). Which statement is false?

A car accelerates uniformly from 5 m/s to 15 m/s in 20 s. It stays at the speed attained at 20 s for 2 min. Finally, the brakes are applied to give a uniform deceleration and it comes to rest in 10 s. Use this data in *Problems 6 to 10*, selecting the correct answer from those given below.

(a) -1.5 m/s^2; (b) $\frac{2}{15}$ m/s^2; (c) 0; (d) 0.5 m/s^2;

(e) 1.3$\dot{8}$ km/h; (f) 7.5 m/s^2; (g) 54 km/h; (h) 2 m/s^2;

(i) 18 km/h; (j) $-\frac{1}{10}$ m/s^2; (k) 1.4$\dot{6}$ km/h; (l) $-\frac{2}{3}$ m/s^2.

6 The initial speed of the car in km/h.

7 The speed of the car after 20 s in km/h.

8 The acceleration during the first 20 s period.

9 The acceleration during the 2 min period.

10 The acceleration during the final 10 s.

(c) CONVENTIONAL PROBLEMS

1 A coach increases speed from 4 km/h to 40 km/h at an average acceleration of 0.2 m/s^2. Find the time taken for this increase in speed. [50 s]

2 A ship changes speed from 15 km/h to 20 km/h in 25 min. Determine the average acceleration in m/s^2 of the ship during this time. [9.26 × 10^{-4} m/s^2]

3 A cyclist travelling at 15 km/h changes speed uniformly to 20 km/h in 1 min, maintains this speed for 5 min and then comes to rest uniformly during the next 15 s. Draw a speed-time graph and hence determine the accelerations in m/s^2 (a) during the first minute, (b) for the next 5 minutes and (c) for the last 10 s.
[(a) 0.0231 m/s^2; (b) 0; (c) -0.370 m/s^2]

4 Assuming uniform accelerations between points, draw the speed-time graph for the data given below, and hence determine the accelerations from A to B, B to C, and C to D.

Point	A	B	C	D
Speed (m/s)	25	5	30	15
Time (s)	15	25	35	45

[A to B -2 m/s^2; B to C 2.5 m/s^2; C to D -1.5 m/s^2]

5 An object is dropped from the third floor of a building. Find its approximate velocity 1.25 s later if all forces except that of gravity are neglected.
[12.25 m/s]

6 During free fall, a ball is dropped from point A and is travelling at 100 m/s when it passes point B. Calculate the time for the ball to travel from A to B if all forces except that of gravity are neglected. [10.2 s]

7 A piston moves at 10 m/s at the centre of its motion and decelerates uniformly at 0.8 m/s^2. Determine its velocity 3 s after passing the centre of its motion.
[7.6 m/s]

8 The final velocity of a train after applying its brakes for 1.2 min is 24 km/h. If its uniform retardation is 0.06 m/s^2, find its speed before the brakes are applied.
[39.6 km/h]

9 A plane in level flight at 400 km/h starts to descend at a uniform acceleration of 0.6 m/s². It levels off when its speed is 670 km/h. Calculate the time during which it is losing height.   [2 min 5 s]

10 A lift accelerates from rest uniformly at 0.9 m/s² for 1.5 s, travels at constant speed for 7 s and then comes to rest in 3 s. Determine its velocity when travelling at constant speed and its acceleration during the final 3 s of its travel.
[1.35 m/s; −0.45 m/s²]

9 Friction

A. MAIN POINTS CONCERNING FRICTION

1 When an object, such as a block of wood, is placed on a floor and sufficient force is applied to the block, the force being parallel to the floor, the block slides across the floor. When the force is removed, motion of the block stops; thus there is a force which resists sliding. This force is called **dynamic** or **sliding friction**. A force may be applied to the block which is insufficient to move it. In this case, the force resisting motion is called the **static friction** or **striction**. Thus there are two categories into which a frictional force may be split:
   (i) dynamic or sliding friction force which occurs when motion is taking place, and
   (ii) static friction force which occurs before motion takes place.

2 There are three factors which affect the size and direction of frictional forces.
   (i) The size of the frictional force depends on the type of surface, (a block of wood slides more easily on a polished metal surface than on a rough concrete surface).
   (ii) The size of the frictional force depends on the size of the force acting at right angles to the surfaces in contact, called the **normal force**. Thus, if the weight of a block of wood is doubled, the frictional force is doubled when it is sliding on the same surface.
   (iii) The direction of the frictional force is always opposite to the direction of motion. Thus the frictional force opposes motion, as shown in *Fig 1*.

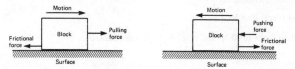

Fig 1

3 The coefficient of friction, μ, is a measure of the amount of friction existing between two surfaces. A low value of coefficient of friction indicates that the force required for sliding to occur is less than the force required when the coefficien of friction is high. The value of the coefficient of friction is given by

$$\mu = \frac{\text{frictional force, } (F)}{\text{normal force, } (N)}$$

Transposing, gives: frictional force = μ × normal force,

$F = \mu N$

(See *Problems 1 to 3*)

The direction of the forces given in this equation are as shown in *Fig 2*. The coefficient of friction is the ratio of a force to a force, and hence has no units. Typical values for the coefficient of friction when sliding is occurring, i.e., the dynamic coefficient of friction are:

For polished oiled metal surfaces	less than 0.1
For glass on glass	0.4
For rubber on tarmac	close to 1.0

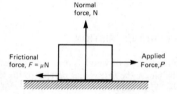

Fig 2

4 In some applications, a low coefficient of friction is desirable, for example, in bearings, pistons moving within cylinders, on ski runs, and so on. However, for such applications as force being transmitted by belt drives and braking systems, a high value of coefficient is necessary. (See *Problems 4 and 5*)

B. WORKED PROBLEMS ON FRICTION

Problem 1 A block of steel requires a force of 10.4 N applied parallel to a steel plate to keep it moving with constant velocity across the plate. If the normal force between the block and the plate is 40 N, determine the dynamic coefficient of friction.

As the block is moving at constant velocity, the force applied must be that required to overcome frictional forces, i.e.,

frictional force, $F = 10.4$ N

The normal force is 40 N, and since $F = \mu N$, (see para. 3),

$\mu = \dfrac{F}{N} = \dfrac{10.4}{40} = 0.26$

i.e. **the dynamic coefficient of friction is 0.26.**

Problem 2 The surface between the steel block and plate of *Problem 1* is now lubricated and the dynamic coefficient of friction falls to 0.12. Find the new value of force required to push the block at a constant speed.

The normal force depends on the weight of the block and remains unaltered at 40 N. The new value of the dynamic coefficient of friction is 0.12 and since the frictional force $F = \mu N$, $F = 0.12 \times 40 = 4.8$ N. The block is sliding at constant speed, thus the force required to overcome the frictional force is also 4.8 N, i.e.,

the required applied force is 4.8 N

Problem 3 The material of a brake is being tested and it is found that the dynamic coefficient of friction between the material and steel is 0.91. Calculate the normal force when the frictional force is 0.728 kN.

The dynamic coefficient of friction, $\mu = 0.91$

The frictional force, $F = 0.728$ kN $= 728$ N

Since $F = \mu N$, then $N = \dfrac{F}{\mu}$

i.e. normal force, $N = \dfrac{728}{0.91} = 800$ N

i.e. **the normal force is 800 N**

Problem 4 State three advantages and three disadvantages of frictional forces.

Instances where frictional forces are an advantage include:
(i) Almost all fastening devices rely on frictional forces to keep them in place once secured, examples being screws, nails, nuts, clips and clamps.
(ii) Satisfactory operation of brakes and clutches rely on frictional forces being present.
(iii) In the absence of frictional forces, most accelerations along a horizontal surface are impossible. For example, a person's shoes just slip when walking is attempted and the tyres of a car just rotate with no forward motion of the car being experienced.

Disadvantages of frictional forces include:
(i) Energy is wasted in the bearings associated with shafts, axles and gears due to heat being generated.
(ii) Wear is caused by friction, for example, in shoes, brake lining materials and bearings.
(iii) Energy is wasted when motion through air occurs, (it is much easier to cycle with the wind rather than against it).

Problem 5 Discuss briefly two design implications which arise due to frictional forces and how lubrication may or may not help.

(i) Bearings are made of an alloy called white metal, which has a relatively low melting point. When the rotating shaft rubs on the white metal bearing, heat is generated by friction, often in one spot and the white metal may melt in this area, rendering the bearing useless. Adequate lubrication, (oil or grease), separates the shaft from the white metal, keeps the coefficient of friction small and prevents damage to the bearing. For very large bearings, oil is pumped under pressure into the bearing and the oil is used to remove the heat generated, often passing through oil coolers before being recirculated. Designers should ensure that the heat generated by friction can be dissipated.
(ii) Wheels driving belts, to transmit force from one place to another, are used in many workshops. The coefficient of friction between the wheel and the belt

must be high, and it may be increased by dressing the belt with a tar-like substance. Since frictional force is proportional to the normal force, a slipping belt is made more efficient by tightening it, thus increasing the normal and hence the frictional force. Designers should incorporate some belt tension mechanism into the design of such a system.

Problem 6 Explain what is meant by the terms (a) the limiting or static coefficient of friction and (b) the sliding or dynamic coefficient of friction.

(a) When an object is placed on a surface and a force is applied to it in a direction parallel to the surface, if no movement takes place, then the applied force is balanced exactly by the frictional force. As the size of the applied force is increased, a value is reached such that the object is just on the point of moving. The limiting or static coefficient of friction is given by the ratio of this applied force to the normal force, where the normal force is the force acting at right angles to the surfaces in contact.

(b) Once the applied force is sufficient to overcome the striction its value can be reduced slightly and the object moves across the surface. A particular value of the applied force is then sufficient to keep the object moving at a constant velocity. The sliding or dynamic coefficient of friction is the ratio of the applied force, to maintain constant velocity, to the normal force.

C. FURTHER PROBLEMS ON FRICTION

(a) SHORT ANSWER PROBLEMS

1 The .... of frictional force depends on the .... of surfaces in contact.

2 The .... of frictional force depends on the size of the .... to the surfaces in contact.

3 The .... of frictional force is always .... to the direction of motion.

4 The coefficient of friction between surfaces should be a .... value for materials concerned with bearings.

5 The coefficient of friction should have a .... value for materials concerned with braking systems.

6 The coefficient of dynamic or sliding friction is given by $\frac{............}{............}$.

7 The coefficient of static or limiting friction is given by $\frac{............}{............}$, when .... is just about to take place.

8 Lubricating surfaces in contact results in a .... of the coefficient of friction.

(b) MULTI-CHOICE PROBLEMS (answers on page 148)

Problems 1 to 5 refer to the statements given below. Select the statement required from each group given.

(a) The coefficient of friction depends on the type of surfaces in contact.

(b) The coefficient of friction depends on the force acting at right angles to the surfaces in contact.

(c) The coefficient of friction depends on the area of the surfaces in contact.

(d) Frictional force acts in the opposite direction to the direction of motion.

(e) Frictional force acts in the direction of motion.

(f) A low value of coefficient of friction is required between the belt and the wheel in a belt drive system.

(g) A low value of coefficient of friction is required for the materials of a bearing.

(h) The dynamic coefficient of friction is given by $\dfrac{\text{normal force}}{\text{frictional force}}$ at constant speed.

(i) The coefficient of static friction is given by $\dfrac{\text{applied force}}{\text{frictional force}}$ as sliding is just about to start.

(j) Lubrication results in a reduction in the coefficient of friction.

1 Which statement is false from (a), (b), (f) and (i)?
2 Which statement is false from (b), (e), (g) and (j)?
3 Which statement is true from (c), (f), (h) and (i)?
4 Which statement is false from (b), (c), (e) and (j)?
5 Which statement is false from (a), (d), (g) and (h)?

6 The normal force between two surfaces is 100 N and the dynamic coefficient of friction is 0.4. The force required to maintain a constant speed of sliding is:
(a) 100.4 N; (b) 40 N; (c) 99.6 N; (d) 250 N.

7 The normal force between two surfaces is 50 N and the force required to maintain a constant speed of sliding is 25 N. The dynamic coefficient of friction is:
(a) 25; (b) 2; (c) 75; (d) 0.5.

8 The maximum force which can be applied to an object without sliding occurring is 60 N, and the static coefficient of friction is 0.3. The normal force between the two surfaces is:
(a) 200 N; (b) 18 N; (c) 60.3 N; (d) 59.7 N.

(c) CONVENTIONAL PROBLEMS

1 Briefly discuss the factors affecting the size and direction of frictional forces.

2 Name three practical applications where a low value of coefficient of friction is desirable and state briefly how this is achieved in each case.

3 Give three practical applications where a high value of coefficient of friction is required when transmitting forces and discuss how this is achieved.

4 For an object on a surface, two different values of coefficient of friction are possible. Give the names of these two coefficients of friction and state how their values may be obtained.

5 Discuss briefly the effects of frictional force on the design of (a) a hovercraft, (b) a screw and (c) a braking system.

6 The coefficient of friction of a brake pad and a steel disc is 0.82. Determine the normal force between the pad and the disc if the frictional force required is 1025 N. [1250 N]

7. A force of 0.12 kN is needed to push a bale of cloth along a chute at a constant speed. If the normal force between the bale and the chute is 500 N, determine the dynamic coefficient of friction. [0.25]

8. The normal force between a belt and its driver wheel is 750 N. If the static coefficient of friction is 0.9 and the dynamic coefficient of friction is 0.87, calculate (a) the maximum force which can be transmitted and (b) maximum force which can be transmitted when the belt is running at a constant speed.
[(a) 675 N; (b) 652.5 N]

10 Waves

A. MAIN POINTS CONCERNED WITH WAVES

1. **Wave motion** is a travelling disturbance through a medium or through space, in which energy is transferred from one point to another without movement of matter.
2. **Examples where wave motion occurs** include:
    (i) Water waves, such as are produced when a stone is thrown into a still pool of water;
    (ii) waves on strings;
    (iii) waves on stretched springs;
    (iv) sound waves;
    (v) light waves (see chapter 11);
    (vi) radio waves;
    (vii) infra-red waves, which are emitted by hot bodies;
    (viii) ultra-violet waves, which are emitted by very hot bodies and some gas discharge lamps;
    (ix) x-ray waves, which are emitted by metals when they are bombarded by high speed electrons;
    (x) gamma-rays, which are emitted by radioactive elements.
    Examples (i) to (iv) are **mechanical waves** and they require a medium (such as air or water) in order to move. Examples (v) to (x) are **electromagnetic waves** and do not require any medium—they can pass through a vacuum.
3. There are two types of wave, these being transverse and longitudinal waves.
    (i) **Transverse waves** are where the particles of the medium move perpendicular to the direction of movement. For example, when a stone is thrown into a pool of still water, the ripple moves radially outwards but the movement of a floating object shows that the water at a particular point merely moves up and down. Light and radio waves are other examples of transverse waves.
    (ii) **Longitudinal waves** are where the particles of the medium vibrate back and forth parallel to the direction of the wave travel. Examples include sound waves and waves in springs.
4. *Fig 1* shows a cross-section of a typical wave.
    (i) **Wavelength** is the distance between two successive identical parts of a wave (for example, between two crests as shown in *Fig 1*).
    The symbol for wavelength is λ (Greek lambda) and its unit is metres.

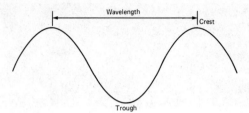

Fig 1

(ii) **Frequency** is the number of complete waves (or cycles) passing a fixed point in one second.

The symbol for frequency is f and its unit is the hertz, Hz.

(iii) The **velocity**, v, of a wave is given by:

velocity = frequency × wavelength

i.e. $v = f\lambda$

The unit of velocity is metres per second.

(See *Problems 1 to 4*)

5 **Reflection** is a change in direction of a wave while the wave remains in the same medium. There is no change in the speed of a reflected wave. All waves are reflected when they meet a surface through which they cannot pass. For example,
   (i) light waves are reflected by mirrors;
   (ii) water waves are reflected at the end of a bath or by a sea wall;
   (iii) sound waves are reflected at a wall (which can produce an echo);
   (iv) a wave reaching the end of a spring or string is reflected; and
   (v) television waves are reflected by satellites above the Earth.
   Experimentally, waves produced in an open tank of water may readily be observed to reflect off a sheet of glass placed at right angles to the surface of the water.

6 **Refraction** is a change in direction of a wave as it passes from one medium to another. All waves refract, and examples include:
   (i) a light wave changing its direction at the boundary between air and glass (see *Fig 2*);
   (ii) sea waves refracting when reaching more shallow waters, and
   (iii) sound waves refracting when entering air of different temperature (see para. 8).

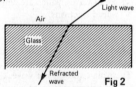

Fig 2

Experimentally, if one end of a water tank is made shallow the waves may be observed to travel more slowly in these regions and are seen to change direction as the wave strikes the boundary of the shallow area. The greater the change of velocity the greater is the bending or refraction.

7 A **sound wave** is a series of alternate layers of air, one layer at a pressure slightly higher than atmospheric, called compressions, and the other slightly lower, called rarefactions. In other words, **sound is a pressure wave.** *Fig 3(a)* represents layers of undisturbed air. *Fig 3(b)* shows what happens to the air when a sound wave passe

8 **Characteristics of sound waves**
   (i) Sound waves can travel through solids, liquids and gases, but not through a vacuum.
   (ii) Sound has a finite (i.e. fixed) velocity, the value of which depends on the

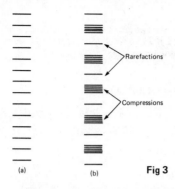

Fig 3

medium through which it is travelling. The velocity of sound is also affected by temperature. Some typical values for the velocity of sound are:
air 331 m/s at 0°C, and 342 m/s at 18°C, water 1410 m/s at 20°C and iron 5100 m/s at 20°C.

(iii) Sound waves can be reflected, the most common example being an echo. Echo-sounding is used for charting the depth of the sea.
(iv) Sound waves can be refracted. This occurs, for example, when sound waves meet layers of air at different temperatures. If a sound wave enters a region of higher temperature the medium has different properties and the wave is bent as shown in *Fig 4*, which is typical of conditions that occur at night.

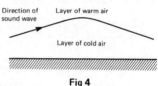

Fig 4

9 Sound waves are produced as a result of vibrations.
   (i) In brass instruments, such as trumpets and trombones, or wind instruments, such as clarinets and oboes, sound is due to the vibration of columns of air.
   (ii) In stringed instruments, such as guitars and violins, sound is produced by vibrating strings causing air to vibrate. Similarly, the vibration of vocal chords produces speech.
   (iii) Sound is produced by a tuning fork due to the vibration of the metal prongs.
   (iv) Sound is produced in a loudspeaker due to vibrations in the cone.
10 The pitch of a sound depends on the frequency of the vibration; the higher the frequency, the higher is the pitch. The frequency of sound depends on the form of the vibrator. The valves of a trumpet or the slide of a trombone lengthen or shorten the air column and the fingers alter the length of strings on a guitar or violin. The shorter the air column or vibrating string the higher the frequency and hence pitch. Similarly, a short tuning fork will produce a higher pitch note than a long tuning fork. Frequencies between about 20 Hz and 20 kHz can be perceived by the human ear.

B. WORKED PROBLEMS ON WAVES

Problem 1 BBC radio 4 is transmitted at a frequency of 200 kHz and a wavelength of 1500 m. Determine the velocity of radio waves.

Velocity = frequency × wavelength
= 200×10^3 Hz × 1500 m = 3×10^8 m/s.

Problem 2 Calculate the wavelength of a sound wave of frequency 2 kHz if the velocity of sound in air is 340 m/s.

Since velocity, $v = f\lambda$ then wavelength, $\lambda = \dfrac{v}{f}$

Hence wavelength, $\lambda = \dfrac{340 \text{ m/s}}{2 \times 10^3} = 0.17$ m.

Problem 3 Water waves of wavelength 15 mm are found to travel 1200 mm in 2 s. Determine their velocity and frequency.

Velocity of wave = $\dfrac{\text{distance travelled}}{\text{time taken}} = \dfrac{1200 \text{ mm}}{2 \text{ s}} = \dfrac{1.2 \text{ m}}{2 \text{ s}} = 0.6$ m/s

Since velocity, $v = f\lambda$, frequency, $f = \dfrac{v}{\lambda} = \dfrac{0.6 \text{ m/s}}{15 \times 10^{-3} \text{ m}} = \dfrac{0.6}{0.015} = 40$ Hz.

Problem 4 A local radio station transmits programmes at a wavelength of 300 m. If the velocity of radio waves is 3×10^8 m/s, determine the frequency of the waves.

Since velocity, $v = f\lambda$, then frequency, $f = \dfrac{v}{\lambda}$.

Hence, frequency, $f = \dfrac{3 \times 10^8 \text{ m/s}}{300 \text{ m}} = 1 \times 10^6$ Hz or **1 MHz**.

C. FURTHER PROBLEMS ON WAVES

(a) SHORT ANSWER PROBLEMS

1 List five examples where wave motion can occur.

2 State the two types of wave, and give two examples of each type.

3 Explain using a simple diagram the meaning of (a) wavelength and (b) frequency.

4 What is the unit of frequency?

5 The velocity v of a wave of frequency f and wavelength λ is given by
   $v = $ ....

6 What is meant by the reflection of a wave? Give three examples of wave reflection.

7 Explain briefly how the reflection of a wave can be demonstrated experimentally.

8 What is meant by the refraction of a wave? Give two examples of wave refraction.

9 Explain briefly how the refraction of a wave can be demonstrated experimentally.

10 Briefly explain what a sound wave is.

11 State three characteristics of sound waves.

12 How is the frequency of a sound wave changed in (a) a brass instrument, and (b) a stringed instrument?

(b) MULTI-CHOICE PROBLEMS (answers on page 149)

1 Waves having a wavelength of 100 mm are found experimentally to travel 10 m in 10 s. The velocity of the waves is:
  (a) 100 m/s; (b) 1000 m/s; (c) 1 m/s; (d) 0.1 m/s.

2 For the waves in *Problem 1*, the frequency is:
  (a) 1 kHz; (b) 10 Hz; (c) 10 kHz; (d) 1 Hz.

3 Which of the following statements is false?
  (a) A light wave is an example of a transverse wave.
  (b) A sound wave is an example of a longitudinal wave.
  (c) The value of the velocity of sound depends on the medium and the temperature.
  (d) Refraction is a change of direction of a wave while the wave remains in the same medium.

4 A radio programme is transmitted at a frequency of 1200 kHz. If the velocity of radio waves is 3×10^8 m/s, the wavelength is:
  (a) 250 m; (b) 4 m; (c) 3.6×10^{14} m; (d) 360 m.

5 Which of the following statements is true?
  (a) As a trombone slide is lengthened the frequency of the sound produced is increased.
  (b) Sound waves are able to travel through any medium.
  (c) Television programmes are transmitted as waves requiring no medium.
  (d) Echo-sounding depends on the ability of sound to be refracted by the sea bed.

(c) CONVENTIONAL PROBLEMS

1 Determine the frequency of sea waves which are travelling at 30 m/s with a distance between successive troughs of 50 m. [0.6 Hz]

2 BBC radio 3 is transmitted at a frequency of 1215 kHz and a wavelength of 247 m. Determine the velocity of the radio waves. [3×10^8 m/s]

3 The wavelength of a sound wave is 100 mm and its frequency is 3.4 kHz. Calculate the velocity of the sound wave. [340 m/s]

4 A radio station transmits programes at a frequency of 95 MHz. If the velocity of radio waves is 3×10^8 m/s, determine the wavelength of the wave. [3.16 m]

5 Water waves of wavelength 25 mm are found to travel 150 cm in 3 s. Determine their velocity and frequency. [0.5 m/s; 20 Hz]

6 A vocalist sings a note having a frequency of 512 Hz. Determine the wavelength of this note if the velocity of sound in air is 340 m/s. [0.664 m]

7 Explain the difference between reflection and refraction of waves and give two examples of each.

8 What is the difference between transverse and longitudinal waves? State two examples of each type.

9 State four examples of how sound may be produced and explain how the pitch of a note may be varied.

11 Light rays

A. MAIN POINTS CONCERNED WITH LIGHT RAYS

1. (i) Light is an electromagnetic wave (see chapter 10) and the straight line paths followed by very narrow beams of light, along which light energy travels, are called **rays**.
   (ii) The behaviour of light rays may be investigated by using a **ray-box**. This consists merely of a lamp in a box containing a narrow slit which emits rays of light.
   (iii) Light always travels in straight lines although its direction can be changed by reflection or refraction.

2. **Reflection of light**
   Fig 1 shows a ray of light, called the incident ray, striking a plane mirror at 0, and making an angle i with the normal, which is a line drawn at right angles to the mirror at 0. i is called the **angle of incidence**. The light ray reflects as shown making an angle r with the normal. r is called the **angle of reflection**.

   There are two **laws of reflection**:
   (i) The angle of incidence is equal to the angle of reflection (i.e. $i = r$ in *Fig 1*).
   (ii) The incident ray, the normal at the point of incidence and the reflected ray all lie in the same plane.

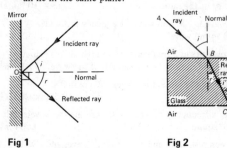

Fig 1   Fig 2

3. **Refraction of light**
   (i) When a ray of light passes from one medium to another the light undergoes a change in direction. This displacement of light rays is called **refraction**.
   (ii) *Fig 2* shows the path of a ray of light as it passes through a parallel sided

74

glass block. The incident ray AB which has an angle of incidence *i* enters the glass block at B. The direction of the ray changes to BC such that the angle *r* is less than angle *i*. *r* is called the angle of refraction. When the ray emerges from the glass at C the direction changes to CD, angle *r'* being greater than *i'*. The final emerging ray CD is parallel to the incident ray AB.

(iii) In general, when entering a more dense medium from a less dense medium, light is refracted towards the normal and when it passes from a dense to a less dense medium it is refracted away from the normal.

4 (i) **Lenses** are pieces of glass or other transparent material with spherical surface on one or both sides. When light is passed through a lens it is refracted.

(ii) Lenses are used in spectacles, magnifying glasses and microscopes, telescopes, cameras and projectors.

(iii) There are a number of different shaped lenses and two of the most common are shown in *Fig 3*. *Fig 3(a)* shows a **bi-convex lens**, so called since both its surfaces curve outwards. *Fig 3(b)* shows a **bi-concave lens**, so called since both of its surfaces curve inwards. The line passing through the centre of curvature of the lens surface is called the **principal axis**.

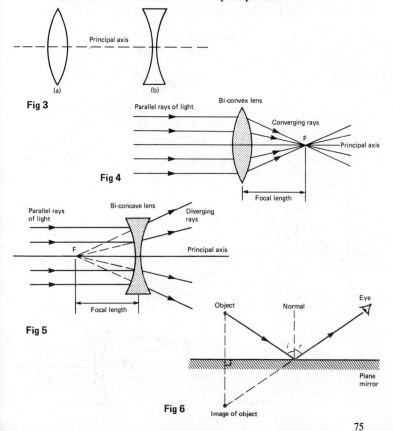

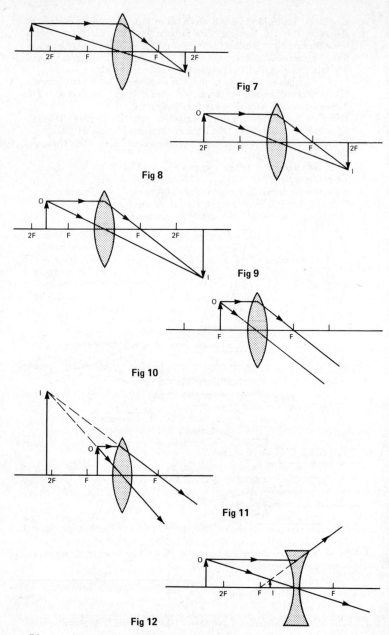

Fig 7

Fig 8

Fig 9

Fig 10

Fig 11

Fig 12

5. (i) *Fig 4* shows a number of parallel rays of light passing through a bi-convex lens. They are seen to converge at a point F on the principal axis.
   (ii) *Fig 5* shows parallel rays of light passing through a bi-concave lens. They are seen to diverge such that they appear to come from a point F which lies between the source of light and the lens, on the principal axis.
   (iii) In both *Figs 4 and 5*, F is called the **principal focus** or the **focal point**, and the distance from F to the centre of the lens is called the **focal length** of the lens.

6. An **image** is the point from which reflected rays of light entering the eye appear to have originated. If the rays actually pass through the point then a **real image** is formed. Such images can be formed on a screen. *Fig 6* illustrates how the eye collects rays from an object after reflection from a plane mirror. To the eye, the rays appear to come from behind the mirror and the eye sees what seems to be an image of the object as far behind the mirror as the object is in front. Such an image is called a **virtual image** and this type cannot be shown on a screen.

7. Lenses are important since they form images when an object emitting light is placed at an appropriate distance from the lens.

   (a) **Bi-convex lenses**

   (i) *Fig 7* shows an object O (a source of light) at a distance of more than twice the focal length from the lens. To determine the position and size of the image two rays only are drawn, one parallel with the principal axis and the other passing through the centre of the lens. The image, I, produced is real, inverted (i.e. upside down), smaller than the object (i.e. diminished) and at a distance between one and two times the focal length from the lens. This arrangement is used in a camera.

   (ii) *Fig 8* shows an object O at a distance of twice the focal length from the lens. The image I is real, inverted, the same size as the object and at a distance of twice the focal length from the lens. This arrangement is used in a photocopier.

   (iii) *Fig 9* shows an object O at a distance of between one and two focal lengths from the lens. The image I is real, inverted, magnified (i.e. greater than the object) and at a distance of more than twice the focal length from the lens. This arrangement is used in a projector.

   (iv) *Fig 10* shows an object O at the focal length of the lens. After passing through the lens the rays are parallel. Thus the image I can be considered as being found at infinity and being real, inverted and very much magnified. This arrangement is used in a spotlight.

   (v) *Fig 11* shows an object O lying inside the focal length of the lens. The image I is virtual, since the rays of light only appear to come from it, is on the same side of the lens as the object, is upright and magnified. This arrangement is used in a magnifying glass.

   (b) **Bi-concave lenses**

   For a bi-concave lens, as shown in *Fig 12*, the object O can be any distance from the lens and the image I formed is virtual, upright, diminished and is found on the same side of the lens as the object. This arrangement is used in some types of spectacles.

B. WORKED PROBLEMS ON LIGHT RAYS

Problem 1 By using plane mirrors show, with an appropriate sketch, how a simple periscope operates.

A simple periscope arrangement is shown in *Fig 13*. A ray of light from O strikes a plane mirror at an angle of 45° at point P. Since from the laws of reflection the angle of incidence i is equal to the angle of reflection r then $i = r = 45°$. Thus angle OPQ = 90° and the light ray is reflected through 90°. The ray then strikes another

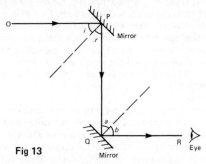

Fig 13

mirror at 45° at point Q. Thus $a = b = 45°$, angle PQR = 90° and the light ray is again reflected through 90°. Thus the light from O finally travels in the direction QR, which is parallel to OP, but displaced by the distance PQ. The arrangement thus acts as a periscope.

Problem 2 A 15 mm high object stands on the principal axis of a bi-convex lens of focal length 25 mm. The object is placed 40 mm from the lens. Determine, by means of a ray diagram the height, form and position of the image formed.

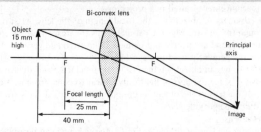

Fig 14

The ray diagram is shown in *Fig 14*. The image is measured as **26 mm high**, is **real, inverted** and lying **70 mm** from the lens on the opposite side to the object.

Problem 3 A 30 mm high object stands on the principal axis of a bi-concave lens. The object is 80 mm from the lens and the focal length of the lens is 40 mm. Determine, by drawing a ray diagram, the height, form and position of the image formed.

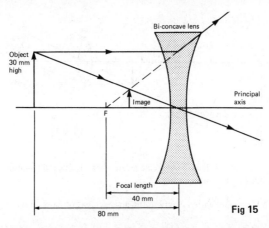

Fig 15

The ray diagram is shown in *Fig 15*. The image is measured as **10 mm high**, is **virtual, upright** and lying **27 mm** from the lens on the same side as the object.

Problem 4 Explain the action of a magnifying glass.

In a magnifying glass a bi-convex lens having a short focal length is used with the object O being placed just inside the focal length as shown in *Fig 16*. The

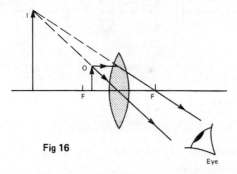

Fig 16

image I is found to lie on the same side of the lens as the object, is virtual, upright and magnified.

Problem 5 Describe the arrangement of simple lenses in a compound microscope, and explain their function.

A compound microscope is able to give large magnification by the use of two (or more) lenses. An object O is placed outside the focal length F_0 of a bi-convex lens, called the objective lens (since it is near to the object), as shown in *Fig 17*. This produces a real, inverted, magnified image I_1. This image then acts as the object for the eyepiece lens, (i.e., the lens nearest the eye), and falls inside the

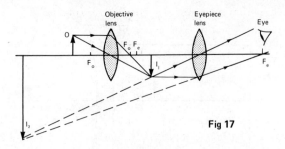

Fig 17

focal length F_e of this lens. The eyepiece lens then produces a magnified, virtual, inverted image I_2 as shown in *Fig 17*.

Problem 6 Describe, with a diagram, the arrangement of lenses in a simple projector.

A simple projector arrangement is shown in *Fig 18* and consists of a source of light and two lens systems. L is a brilliant source of light, such as a tungsten filament. One lens system called the **condenser**, (usually consisting of two

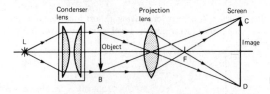

Fig 18

converging lenses as shown), is used to produce an intense illumination of the object AB, which is a slide transparency or film. The second lens, called the **projection lens**, is used to form a magnified, real, upright image of the illuminated object on a distant screen CD.

C. FURTHER PROBLEMS ON LIGHT RAYS

(a) SHORT ANSWER PROBLEMS

1 State the laws of reflection of light rays.

2 Draw the path of a ray of light passing through a parallel-sided glass block.

3 What is meant by the refraction of light?

4 Sketch (a) a bi-concave lens, and (b) a bi-convex lens.

5 What is meant by the focal length of a lens system.

6 Explain the difference between a real and a virtual image.

7 Briefly describe the action of a magnifying-glass.

8 State five practical uses of lenses.

9 Explain briefly how lenses are used in a compound microscope.

10 In the following sentence delete as appropriate:
For an object placed at a distance of two focal lengths from a bi-concave lens the resulting image is: real/virtual, upright/inverted and magnified/diminished.

(b) MULTI-CHOICE PROBLEMS (answers on page 149)

1 Which of the following statements is false?
An object is placed at a distance of one and a half focal lengths from a bi-convex lens. The image produced is:
(a) more than two focal lengths from the lens;
(b) real;
(c) upright;
(d) magnified.

2 Which of the following statements is true?
An object is placed at a distance of three-quarters of the focal length from a bi-concave lens. The image produced is:
(a) inverted;
(b) magnified;
(c) on the opposite side of the lens to the object;
(d) virtual.

3 A beam of light strikes a plane mirror at an angle of incidence of 25°. The angle the reflected ray makes with the mirror is:
(a) 25°; (b) 65°; (c) 50°; (d) 130°.

4 Which of the following statements is false?
(a) When light enters a more dense medium from a less dense medium, it is bent away from the normal.
(b) When light is reflected at a plane mirror, the angle of incidence is equal to the angle of reflection.
(c) The distance from the principal focus to the centre of a lens is called the focal length of the lens.
(d) Some spectacles are made with bi-concave lenses.

5 A ray of light strikes a mirror at an angle of incidence of 35°. The angle between the incident and reflected ray is:
(a) 35°; (b) 55°; (c) 130°; (d) 70°.

(c) CONVENTIONAL PROBLEMS

1 An object, 20 mm high, stands on the principal axis of a bi-convex lens. The object is 50 mm from the lens and the focal length of the lens is 30 mm. Determine the height, form and position of the image formed by drawing a ray diagram.
[30 mm; inverted; 75 mm from lens]

2 If the bi-convex lens of *Problem 1* is replaced by a bi-concave lens having the same focal length, determine the new height, form and position of the image formed.
[7 mm; upright; 18.5 mm from lens]

3 If a light ray strikes a mirror at an angle of incidence of 32°, determine (a) the angle of reflection, and (b) the angle the reflected ray makes with the mirror.
[(a) 32°; (b) 58°]

4 State the laws of reflection. Show by a diagram how the laws of reflection may be used with four plane mirrors to 'see' through an object.

5 Draw a diagram to show the path of a light ray which strikes a plane mirror at an angle of 50°. What is: (a) the angle of incidence, and (b) the angle of reflection.

[(a) 40°; (b) 40°]

6 Explain why a straight rod standing partly in and partly out of water appears to be bent when viewed from an angle.

7 Draw diagrams to illustrate (a) a bi-convex lens, and (b) a bi-concave lens and explain how each type of lens affects light rays.

12 Work, energy and power

A. MAIN POINTS CONCERNED WITH WORK, ENERGY AND POWER

1. Fuel, such as oil, coal, gas or petrol, when burnt, produces heat. Heat is a form of energy and may be used, for example, to boil water or to raise steam. Thus fuel is useful since it is a convenient method of storing energy, that is, **fuel is a source of energy.**

2. (i) If a body moves as a result of a force being applied to it, the force is said to do work on the body. The amount of work done is the product of the applied force and the distance, i.e.
   Work done = force × distance moved in the direction of the force
   (ii) The unit of work is the **joule, J,** which is defined as the amount of work done when a force of 1 Newton acts for a distance of 1 m in the direction of the force.

   Thus, 1 J = 1 Nm.

3. If a graph is plotted of experimental values of force (on the vertical axis) against distance moved (on the horizontal axis) a force-distance graph or work diagram is produced. **The area under the graph represents the work done.**

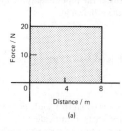 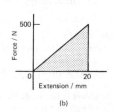

Fig 1

For example, a constant force of 20 N used to raise a load a height of 8 m may be represented on a force-distance graph as shown in *Fig 1(a)*. The area under the graph shown shaded, represents the work done.

Hence, work done = 20 N × 8 m = **160 J.**

Similarly, a spring extended by 20 mm by a force of 500 N may be represented by the work diagram shown in *Fig 1(b)*.

83

Work done = shaded area = $\frac{1}{2} \times$ base $\times$ height = $\frac{1}{2} \times (20 \times 10^{-3})$ m $\times$ 500 N = **5 J**.

(See *Problems 1 to 5*)

4 **Energy** is the capacity, or ability, to do work. The unit of energy is the joule, the same as for work. Energy is expended when work is done.

5 There are several **forms of energy** and these include:
   (i) Mechanical energy
   (ii) Heat or thermal energy
   (iii) Electrical energy
   (iv) Chemical energy
   (v) Nuclear energy
   (vi) Light energy
   (vii) Sound energy

6 Energy may be converted from one form to another. **The principle of conservation of energy** states that the total amount of energy remains the same in such conversions, i.e., energy cannot be created or destroyed. Some examples of energy conversions include:
   (i) Mechanical energy is converted to electrical energy by a generator.
   (ii) Electrical energy is converted to mechanical energy by a motor.
   (iii) Heat energy is converted to mechanical energy by a steam engine.
   (iv) Mechanical energy is converted to heat energy by friction.
   (v) Heat energy is converted to electrical energy by a solar cell.
   (vi) Electrical energy is converted to heat energy by an electric fire.
   (vii) Heat energy is converted to chemical energy by living plants.
   (viii) Chemical energy is converted to heat energy by burning fuels.
   (ix) Heat energy is converted to electrical energy by a thermocouple.
   (x) Chemical energy is converted to electrical energy by batteries.
   (xi) Electrical energy is converted to light energy by a light bulb.
   (xii) Sound energy is converted to electrical energy by a microphone.
   (xiii) Electrical energy is converted to chemical energy by electrolysis.

7 **Efficiency** is defined as the ratio of the useful output energy to the input energy. The symbol for efficiency is η (Greek letter eta).

   Hence, efficiency, $\eta = \dfrac{\text{useful output energy}}{\text{input energy}}$

   Efficiency has no units and is often stated as a percentage. A perfect machine would have an efficiency of 100%. However, all machines have an efficiency lower than this due to friction and other losses. Thus, if the input energy to a motor is 1000 J and the output energy is 800 J then the efficiency is

   $\dfrac{800}{1000} \times 100\%$, i.e. 80%.

   (See *Problems 6 to 9*)

8 **Power** is a measure of the rate at which work is done or at which energy is converted from one form to another.

   Power $P = \dfrac{\text{energy used}}{\text{time taken}}$ (or $P = \dfrac{\text{work done}}{\text{time taken}}$)

   The unit of power is the **watt, W,** where 1 watt is equal to 1 joule per second. The watt is a small unit for many purposes and a larger unit called the kilowatt, kW, is used, where 1 kW = 1000 W. The power output of a motor which does 120 kJ of work in 30 s is thus given by

   $P = \dfrac{120 \text{ kJ}}{30 \text{ s}} = 4 \text{ kW}$

   (For electrical power, see page 110).
   (See *Problems 10 to 15*)

B. WORKED PROBLEMS ON WORK, ENERGY AND POWER

Problem 1 Calculate the work done when a force of 40 N pushes an object a distance of 500 m in the same direction as the force.

Work done = force × distance moved in the direction of the force
= 40 N × 500 m = 20 000 J (since 1 J = 1 Nm)

i.e. **work done = 20 kJ**.

Problem 2 Calculate the work done when a mass is lifted vertically by a crane to a height of 5 m, the force required to lift the mass being 98 N.

Work is done in lifting then:

Work done = (weight of the body) × (vertical distance moved)

Weight is the downward force due to the mass of an object.

Hence work done = 98 N × 5 m = **490 J**.

Problem 3 A motor supplies a constant force of 1 kN which is used to move a load a distance of 5 m. The force is then changed to a constant 500 N and the load is moved a further 15 m. Draw the force-distance graph for the operation and from the graph determine the total work done by the motor.

The force-distance graph or work diagram is shown in *Fig 2*. Between points A and B a constant force of 1000 N moves the load 5 m. Between points C and D a constant force of 500 N moves the load from 5 m to 20 m.
Total work done
= area under the force-distance graph
= area ABFE + area CDGF
= (1000 N × 5 m) + (500 N × 15 m)
= 5000 J + 7500 J = 12500 J = **12.5 kJ**

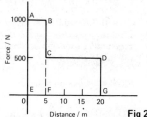

Fig 2

Problem 4 A spring initially in a relaxed state is extended by 100 mm. Determine the work done by using a work diagram if the spring requires a force of 0.6 N per mm of stretch.

Force required for a 100 mm extension
= 100 mm × 0.6 N mm^{-1} = 60 N.

Fig 3 shows the force-extension graph or work diagram representing the increase in extension in proportion to the force, as the force is increased from 0 to 60 N. The work done is the area under the graph (shown shaded).

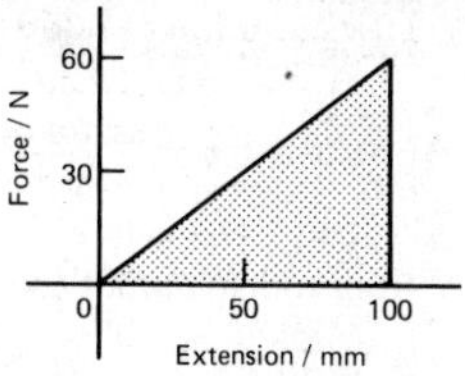

Hence, work done = $\frac{1}{2}$ × base × height = $\frac{1}{2}$ × 100 mm × 60 N

= $\frac{1}{2}$ × 100 × 10^{-3} m × 60 N = **3 J**.

Fig 3

(Alternatively, average force during extension = $\dfrac{60-0}{2}$ = 30 N and

total extension = 100 mm = 0.1 m

Hence work done = average force × extension = 30 N × 0.1 m = **3 J**)

Problem 5 A spring requires a force of 10 N to cause an extension of 50 mm. Determine the work done in extending the spring (a) from zero to 30 mm, and (b) from 30 mm to 50 mm.

Fig 4 shows the force-extension graph for the spring.

(a) Work done in extending the spring from zero to 30 mm is given by area ABO of *Fig 4*, i.e.

Work done = $\dfrac{1}{2}$ × base × height

= $\dfrac{1}{2}$ × 30 × 10^{-3} m × 6 N = 90 × 10^{-3} J

= **0.09 J**

(b) Work done in extending the spring from 30 mm to 50 mm is given by area ABCE of *Fig 4*, i.e.

Work done = area ABCD + area ADE

= (20 × 10^{-3} m × 6 N) + $\dfrac{1}{2}$(20 × 10^{-3} m)(4 N)

= 0.12 J + 0.04 J = **0.16 J**

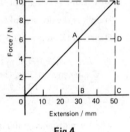

Fig 4

Problem 6 A machine exerts a force of 200 N in lifting a mass through a height of 6 m. If 2 kJ of energy are supplied to it, what is the efficiency of the machine?

Work done in lifting mass = force × distance moved
= weight of body × distance moved
= 200 N × 6 m = 1200 J = useful energy output

Energy output = 2 kJ = 2000 J

Efficiency, $\eta = \dfrac{\text{useful output energy}}{\text{input energy}} = \dfrac{1200}{2000} =$ **0.6 or 60%**

Problem 7 Calculate the useful output energy of an electric motor which is 70% efficient if it uses 600 J of electrical energy.

Efficiency, $\eta = \dfrac{\text{useful output energy}}{\text{input energy}}$, thus $\dfrac{70}{100} = \dfrac{\text{output energy}}{600 \text{ J}}$

from which, output energy = $\dfrac{70}{100}$ × 600 = **420 J**

Problem 8 4 kJ of energy are supplied to a machine used for lifting a mass. The force required is 800 N. If the machine has an efficiency of 50%, to what height will it lift the mass?

Efficiency, $\eta = \dfrac{\text{output energy}}{\text{input energy}}$, i.e. $\dfrac{50}{100} = \dfrac{\text{output energy}}{4000 \text{ J}}$

from which, output energy $= \dfrac{50}{100} \times 4000 = 2000$ J

Work done = force × distance moved
Hence 2000 J = 800 N × height

from which, height $= \dfrac{2000 \text{ J}}{800 \text{ N}} = \mathbf{2.5}$ **m**

Problem 9 A hoist exerts a force of 500 N in raising a load through a height of 20 m. The efficiency of the hoist gears is 75% and the efficiency of the motor is 80%. Calculate the input energy to the hoist.

The hoist system is shown diagrammatically in *Fig 5*.
Output energy = work done = force × distance
$= 500 \text{ N} \times 20 \text{ m} = 10\,000$ J

For the gearing, efficiency $= \dfrac{\text{output energy}}{\text{input energy}}$, i.e. $\dfrac{75}{100} = \dfrac{10\,000}{\text{input energy}}$

from which, the input energy to the gears $= 10\,000 \times \dfrac{100}{75} = 13\,333$ J

Fig 5

The input energy to the gears is the same as the output energy of the motor.

Thus, for the motor, efficiency $= \dfrac{\text{output energy}}{\text{input energy}}$, i.e., $\dfrac{80}{100} = \dfrac{13\,333}{\text{input energy}}$

Hence the input energy to the system $= 13\,333 \times \dfrac{100}{80} = 16\,670$ J $= \mathbf{16.67}$ **kJ**

Problem 10 The output power of a motor is 8 kW. How much work does it do in 30 s?

Power $= \dfrac{\text{work done}}{\text{time taken}}$

From which
work done = power × time
$= 8000 \text{ W} \times 30 \text{ s} = 240\,000 \text{ J} = \mathbf{240}$ **kJ**

Problem 11 Calculate the power required to lift a mass through a height of 10 m in 20 s if the force required is 3924 N.

Work done = force × distance moved = 3924 N × 10 m = 39 240 J

Power $= \dfrac{\text{work done}}{\text{time taken}} = \dfrac{39\,240 \text{ J}}{20 \text{ s}} = \mathbf{1962}$ **W or 1.962 kW**

Problem 12 10 kJ of work is done by a force in moving a body uniformly through 125 m in 50 s. Determine (a) the value of the force and (b) the power.

(a) Work done = force × distance.

Hence 10 000 J = force × 125 m, from which, force = $\dfrac{10\ 000\ \text{J}}{125\ \text{m}}$ = **80 N**

(b) Power = $\dfrac{\text{work done}}{\text{time taken}}$ = $\dfrac{10\ 000\ \text{J}}{50\ \text{s}}$ = **200 W**

Problem 13 A car hauls a trailer at 90 km h^{-1} when exerting a steady pull of 600 N. Calculate (a) the work done in 30 minutes and (b) the power required.

(a) Work done = force × distance moved

Distance moved in 30 mins, i.e., $\dfrac{1}{2}$ h, at 90 km h^{-1} = 45 km

Hence work done = 600 N × 45 000 m = **27 000 kJ or 27 MJ**

(b) Power required = $\dfrac{\text{work done}}{\text{time taken}}$ = $\dfrac{27 \times 10^6\ \text{J}}{30 \times 60\ \text{s}}$ = **15 000 W or 15 kW**

Problem 14 To what height will a mass of weight 981 N be raised in 40 s by a machine using a power of 2 kW?

Work done = force × distance.
Hence, work done = 981 N × height.

Power = $\dfrac{\text{work done}}{\text{time taken}}$, from which, work done = power × time taken

= 2000 W × 40 s = 80 000 J

Hence 80 000 = 981 N × height

from which, height = $\dfrac{80\ 000\ \text{J}}{981\ \text{N}}$ = **81.55 m**

Problem 15 A planing machine has a cutting stroke of 2 m and the stroke takes 4 seconds. If the constant resistance to the cutting tool is 900 N calculate for each cutting stroke (a) the power consumed at the tool point, and (b) the power input to the system if the efficiency of the system is 75%.

(a) Work done in each cutting stroke = force × distance
= 900 N × 2 m = 1800 J.

Power consumed at tool point = $\dfrac{\text{work done}}{\text{time taken}}$ = $\dfrac{1800\ \text{J}}{4\ \text{s}}$ = **450 W**

(b) Efficiency = $\dfrac{\text{output energy}}{\text{input energy}}$ = $\dfrac{\text{output power}}{\text{input power}}$

Hence $\dfrac{75}{100}$ = $\dfrac{450}{\text{input power}}$, from which, input power = 450 × $\dfrac{100}{75}$ = **600 W**

C. FURTHER PROBLEMS ON WORK, ENERGY AND POWER

(a) SHORT ANSWER PROBLEMS

1 Define work in terms of force applied and distance moved.

2 Define energy, and state its unit.

3 Define the joule.

4 The area under a force-distance graph represents ....

5 Name five forms of energy.

6 State the principle of conservation of energy.

7 Give two examples of conversion of heat energy to other forms of energy.

8 Give two examples of conversion of electrical energy to other forms of energy.

9 Give two examples of conversion of chemical energy to other forms of energy.

10 Give two examples of conversion of mechanical energy to other forms of energy.

11 (a) Define efficiency in terms of energy input and energy output.
   (b) State the symbol used for efficiency.

12 Define power and state its unit.

(b) MULTI-CHOICE PROBLEMS (answers on page 149)

1 State which of the following is incorrect.
   (a) $1\text{ W} = 1\text{ J s}^{-1}$;
   (b) $1\text{ J} = 1\text{ N/m}$;
   (c) $\eta = \dfrac{\text{output energy}}{\text{input energy}}$;
   (d) energy = power × time.

2 An object is lifted 2000 mm by a crane. If the force required is 100 N, the work done is:
   (a) $\dfrac{1}{20}$ N; (b) 200 kN; (c) 200 N; (d) 20 N.

3 A motor having an efficiency of 0.8 uses 800 J of electrical energy. The output energy of the motor is (a) 800 J; (b) 1000 J; (c) 640 J.

4 6 kJ of work is done by a force in moving an object uniformly through 120 m in 1 minute. The force applied is
   (a) 50 N; (b) 20 N; (c) 720 N; (d) 12 N.

5 For the object in *Problem 4*, the power developed is:
   (a) 6 kW; (b) 12 kW; (c) $\dfrac{5}{6}$ W; (d) 0.1 kW.

6 Which of the following statements is false?
   (a) The unit of energy and work is the same.
   (b) The area under a force-distance graph gives the work done.
   (c) Electrical energy is converted to mechanical energy by a generator.
   (d) Efficiency is the ratio of the useful output energy to the input energy.

7 A machine using a power of 1 kW requires a force of 100 N to raise a mass in 10 s. The height the mass is raised in this time is:
   (a) 100 m; (b) 1 km; (c) 10 m; (d) 1 m.

8 A force-extension graph for a spring is shown in *Fig 6*. Which of the following statements is false? The work done in extending the spring:
   (a) from 0 to 100 mm is 5 N;

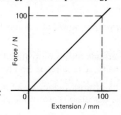

Fig 6

(b) from 0 to 50 mm is 1.25 N;
  (c) from 20 mm to 60 mm is 1.6 N;
  (d) from 60 mm to 100 mm is 3.75 N.

(c) CONVENTIONAL PROBLEMS

1 Determine the work done when a force of 50 N pushes an object 1.5 km in the same direction as the force. [75 kJ]

2 Calculate the work done when a mass of weight 200 N is lifted vertically by a crane to a height of 100 m. [20 kJ]

3 A motor supplies a constant force of 2 kN to move a load 10 m. The force is then changed to a constant 1.5 kN and the load is moved a further 20 m. Draw the force-distance graph for the complete operation, and, from the graph, determine the total work done by the motor. [50 kJ]

4 A spring, initially relaxed, is extended 80 mm. Draw a work diagram and hence determine the work done if the spring requires a force of 0.5 N/mm of stretch. [1.6 J]

5 A spring requires a force of 50 N to cause an extension of 100 mm. Determine the work done in extending the spring (a) from 0 to 100 mm, and (b) from 40 mm to 100 mm. [(a) 2.5 J; (b) 2.1 J]

6 The resistance to a cutting tool varies during the cutting stroke of 800 mm as follows:
   (i) The resistance increases uniformly from an initial 5000 N to 10 000 N as the tool moves 500 mm.
   (ii) The resistance falls uniformly from 10 000 N to 6000 N as the tool moves 300 mm. Draw the work diagram and calculate the work done in one cutting stroke. [6.15 kJ]

7 A machine lifts a mass of weight 490.5 N through a height of 12 m when 7.85 kJ of energy is supplied to it. Determine the efficiency of the machine. [75%]

8 Determine the output energy of an electric motor which is 60% efficient if it uses 2 kJ of electrical energy. [1.2 kJ]

9 State five possible energy conversions for a motor car.

10 A machine which is used for lifting a particular mass is supplied with 5 kJ of energy. If the machine has an efficiency of 65% and exerts a force of 812.5 N to what height will it lift the mass? [4 m]

11 A load is hoisted 42 m and requires a force of 100 N. The efficiency of the hoist gear is 60% and that of the motor is 70%. Determine the input energy to the hoist. [10 kJ]

12 The output power of a motor is 10 kW. How much work does it do in 1 minute? [600 kJ]

13 Determine the power required to lift a load through a height of 20 m in 12.5 s if the force required is 2.5 kN. [4 kW]

14 25 kJ of work is done by a force in moving an object uniformly through 50 m in 40 s. Calculate (a) the value of the force, and (b) the power.
[(a) 500 N; (b) 625 W]

15 A car towing another at 54 km/h exerts a steady pull of 800 N. Determine (a) the work done in 1/4 hr, and (b) the power required.

[(a) 10.8 MJ; (b) 12 kW]

16 To what height will a mass of weight 500 N be raised in 20 s by a motor using 4 kW of power? [160 m]

17 The output power of a motor is 10 kW. Determine (a) the work done by the motor in 2 hours, and (b) the energy used by the motor if it is 72% efficient.

[(a) 72 MJ; (b) 100 MJ]

13 Heat energy

A. MAIN POINTS CONCERNED WITH HEAT ENERGY

1. (i) **Heat** is a form of energy and is measured in joules.
   (ii) **Temperature** is the degree of hotness or coldness of a substance.
   Heat and temperature are thus **not** the same thing. For example, twice the heat energy is needed to boil a full container of water than half a container—that is, different amounts of heat energy are needed to cause an equal rise in the temperature of different amounts of the same substance.

2. Temperature is measured either (i) on the **Celsius (°C) scale** (formerly Centigrade), where the temperature at which ice melts, i.e. the freezing point of water, is taken as 0°C and the point at which water boils under normal atmospheric pressure is taken as 100°C, or (ii) on the **thermodynamic scale**, in which the unit of temperature is the kelvin (K). The kelvin scale uses the same temperature interval as the Celsius scale but as its zero takes the "absolute zero of temperature" which is at about −273°C.
   Hence, kelvin temperature = degree Celsius + 273

   i.e. **K = (°C) + 273**

   Thus, for example, 0°C = 273 K, 25°C = 298 K and 100°C = 373 K.
   (See *Problems 1 and 2*)

3. A **thermometer** is an instrument which measures temperature. Any substance which possesses one or more properties which vary with temperature can be used to measure temperature. These properties include changes in length, area or volume, electrical resistance or in colour. Examples of temperature measuring devices include:
   (i) **liquid-in-glass thermometer**, which uses the expansion of a liquid with increase in temperature as its principle of operation, (see *Problem 3*);
   (ii) **thermocouples**, which use the emf set up when the junction of two dissimilar metals is heated, (see *Problem 4*);
   (iii) **resistance thermometer**, which uses the change in electrical resistance caused by temperature change, and
   (iv) **pyrometers**, which are devices for measuring very high temperatures, using the principle that all substances emit radiant energy when hot, the rate of emission depending on their temperature.

4. (i) The **specific heat capacity** of a substance is the quantity of heat energy required to raise the temperature of 1 kg of the substance by 1°C.

(ii) The symbol used for specific heat capacity is c and the units are J/(kg °C) or J/(kg K). (Note that these units may also be written as J kg^{-1} °C^{-1} or J kg^{-1} K^{-1}.)

(iii) Some typical values of specific heat capacity for the range of temperature 0°C to 100°C include:

Water, 4190 J/(kg °C);   Ice, 2100 J/(kg °C);
Aluminium, 950 J/(kg °C);   Copper, 390 J/(kg °C);
Iron, 500 J/(kg °C);   Lead, 130 J/(kg °C).

Hence to raise the temperature of 1 kg of iron by 1°C requires 500 J of energy, to raise the temperature of 5 kg of iron by 1°C requires (500 × 5) J of energy, and

to raise the temperature of 5 kg of iron by 40°C requires (500 × 5 × 40) J of energy, i.e., 100 kJ.

In general, the quantity of heat energy, Q, required to raise a mass m kg of a substance with a specific heat capacity c J/(kg °C) from temperature t_1 °C to t_2 °C is given by:

$$Q = mc(t_2 - t_1) \text{ joules.}$$

(See *Problems 5 to 9*)

5 A material may exist in any one of three states—solid, liquid or gas. If heat is supplied at a constant rate to some ice initially at, say, −30°C, its temperature rises as shown in *Fig 1*. Initially the temperature increases from −30°C to 0°C as shown by the line AB. It then remains constant at 0°C for the time BC required for the ice to melt into water.

When melting commences the energy gained by continual heating is offset by the energy required for the change of state and the temperature remains constant even though heating is continued. When the ice is completely melted to water, continual heating raises the temperature to 100°C, as shown by CD in *Fig 1*. The water then begins to boil and the temperature again remains constant at 100°C, shown as DE, until all the water has vaporised.

Continual heating raises the temperature of the steam as shown by EF in the region where the steam is termed superheated.

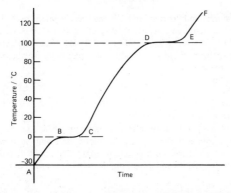

Fig 1

Changes of state from solid to liquid or liquid to gas occur without change of temperature and such changes are reversible processes. When heat energy flows to or from a substance and causes a change of temperature, such as between A and B between C and D and between E and F in *Fig 1*, it is called **sensible heat** (since it can be 'sensed' by a thermometer).

Heat energy which flows to or from a substance while the temperature remains constant, such as between B and C and between D and E in *Fig 1*, is called **latent heat** (latent means concealed or hidden). (See *Problem 10*)

6 (i) The **specific latent heat of fusion** is the heat required to change 1 kg of a substance from the solid state to the liquid state (or vice versa) at constant temperature.
 (ii) The **specific latent heat of vaporisation** is the heat required to change 1 kg of a substance from a liquid to a gaseous state (or vice versa) at constant temperature.
 (iii) The units of the specific latent heats of fusion and vaporisation are J/kg, or more often, kJ/kg, and some typical values are shown in *Table 1*.

TABLE 1

	Latent heat of fusion (kJ/kg)	*Melting point* (°C)
Mercury	11.8	−39
Lead	22	327
Silver	100	957
Ice	335	0
Aluminium	387	660

	Latent heat of vaporisation (kJ/kg)	*Boiling point* (°C)
Oxygen	214	−183
Mercury	286	357
Ethyl alcohol	857	79
Water	2257	100

(iv) The quantity of heat Q supplied or given out during a change of state is given by:

$$Q = mL$$

where m is the mass in kilograms and L is the specific latent heat. Thus, for example, the heat required to convert 10 kg of ice at 0°C to water at 0°C is given by 10 kg × 335 kJ/kg, i.e. 3350 kJ or 3.35 MJ. (See *Problems 11 to 15*)

7 Heat may be transferred from a hot body to a cooler body by one or more of three methods, these being
 (a) by **conduction**;
 (b) by **convection**;
 (c) by **radiation**.
 (See *Problem 16 to 20*)

8 Besides changing temperature, the effects of supplying heat to a material can involve changes in dimensions, as well as in colour, state and electrical resistance. Most substances expand when heated and contract when cooled, and there are many practical applications and design implications of thermal movement. (See *Problem 21*)

B. WORKED PROBLEMS ON HEAT ENERGY

Problem 1 Convert the following temperatures into the kelvin scale:
(a) 37°C; (b) −28°C.

From para. 2, kelvin temperature = degree Celsius + 273.
(a) 37°C corresponds to a kelvin temperature of 37 + 273, i.e., **310 K**.
(b) −28°C corresponds to a kelvin temperature of −28 + 273, i.e., **245 K**.

Problem 2 Convert the following temperatures into the Celsius scale:
(a) 365 K; (b) 213 K.

From para. 2, K = (°C) + 273. Hence, degree Celsius = kelvin temperature −273.
(a) 365 K corresponds to 365−273, i.e. **92°C**.
(b) 213 K corresponds to 213−273, i.e. **−60°C**.

Problem 3 Make a labelled sketch of a liquid-in-glass thermometer and state its principle of operation and its limitations.

A typical liquid-in-glass thermometer is shown in *Fig 2*. When the temperature is increased the liquid (usually either mercury or coloured alcohol) expands and moves up the tube. The position of the end of the column of liquid in the capillary tube is a measure of the temperature of the liquid in the bulb. In *Fig 2* this is

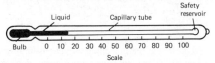

Fig 2

shown at about 15°C—room temperature. In producing a temperature scale, two fixed points are needed. In the Celsius thermometer shown, the fixed points chosen are the temperature of melting ice (0°C) and that of boiling water at normal atmospheric pressure (100°C), in each case the blank stem being marked at the liquid level. The distance between the two points is then divided into 100 equal parts, each equivalent to 1°C, thus forming the scale.

The clinical thermometer, with a limited scale around body temperature, and the maximum and/or minimum thermometer, recording the maximum day temperature and minimum night temperature, are other examples of liquid-in-glass temperature measuring devices.

The choice of mercury as the liquid used has many advantages, for mercury is clearly visible, has a fairly uniform rate of expansion, is readily obtained in the pure state, does not wet the glass and is a good conductor of heat. However mercury has a freezing point of −39°C whereas alcohol has a freezing point of −113°C. Also alcohol is considerably cheaper than mercury.

A disadvantage of liquid-in-glass thermometers in general is that they tend to be fragile.

Problem 4 Describe and explain how a thermocouple can be used with a galvanometer for temperature measurement. Make a labelled sketch of a typical industrial thermocouple and state typical applications, and advantages of such an instrument.

At the junction between two different metals, say, copper and constantan, there exists a difference in electrical potential which varies with the temperature of the junction. If the circuit is completed with a second junction at a different temperature, a current is found to flow round the circuit.

This principle is used in the thermocouple. Two different metal conductors having their ends twisted together are shown in *Fig 3*. If the two junctions X and Y are at different temperatures, a current I flows round the circuit.

Fig 3

The deflection on the galvanometer G depends on the difference in temperature between junctions X and Y and is caused by the difference between the voltage V_x and V_y. The higher temperature junction is usually called the 'hot junction' and the lower temperature junction the 'cold junction'. If the cold junction is kept at a constant known temperature, the galvanometer can be calibrated to indicate the temperature of the hot junction directly.

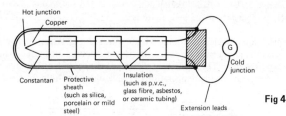

Fig 4

A typical copper–constantan thermocouple suitable for industrial use is shown in *Fig 4*. Such a thermocouple may measure temperatures up to about 400°C and typical applications are found in food processing, boiler flue gas and sub-zero temperature measurement. Iron–constantan thermocouples may measure temperatures up to 850°C and typical applications are found in chemical reactors, paper and pulp mills and re-heat and annealing furnaces. Chromel–alumel thermocouples may measure temperatures up to 1100°C and typical applications are found in blast furnaces, brick kilns and in glass manufacture. Thermocouples made of platinum–rhodium, capable of measuring up to 1400°C and tungsten–molybdenum, capable of measuring up to 2600°C, are also available.

The thermocouple is the most widely used industrial thermometer and its advantages include a very simple, robust construction, that they can be made very small and compact, being easily replaced if damaged, being ideal for remote control situations and there being little delay in the reading becoming steady after a change in temperature.

Problem 5 Calculate the quantity of heat required to raise the temperature of 5 kg of water from 0°C to 100°C. Assume the specific heat capacity of water is 4200 J/(kg °C).

Quantity of heat energy, $Q = mc(t_2 - t_1)$, from para. 4,
$= 5 \text{ kg} \times 4200 \text{ J/(kg °C)} \times (100 - 0)°C$
$= 5 \times 4200 \times 100$
$= 2\ 100\ 000$ J or 2100 kJ or **2.1 MJ**.

Problem 6 A block of cast iron having a mass of 10 kg cools from a temperature of 150°C to 50°C. How much energy is lost by the cast iron? Assume the specific heat capacity of copper is 500 J/(kg °C).

Quantity of heat energy, $Q = mc(t_2 - t_1)$ = 10 kg × 500 J/(kg °C) × (50 − 150)°C
= 10 × 500 × (−100)
= **−500 000 J** or **−500 kJ** or **−0.5 MJ**.

(Note that the minus sign indicates that heat is given out or lost.)

Problem 7 Some lead having a specific heat capacity of 130 J/(kg °C) is heated from 27°C to its melting point at 327°C. If the quantity of heat required is 780 kJ determine the mass of the lead.

Quantity of heat, $Q = mc(t_2 - t_1)$
Hence, 780 × 10³ J = m × 130 J/(kg °C) × (327 − 27)°C
i.e. 780 000 = m × 130 × 300

from which, mass, $m = \dfrac{780\,000}{130 \times 300}$ kg = **20 kg**.

Problem 8 273 kJ of heat energy are required to raise the temperature of 10 kg of copper from 15°C to 85°C. Determine the specific heat capacity of copper.

Quantity of heat, $Q = mc(t_2 - t_1)$
Hence, 273 × 10³ J = 10 kg × c × (85 − 15)°C
where c = specific heat capacity
i.e. 273 000 = 10 × c × 70
from which, specific heat capacity of copper,

$c = \dfrac{273\,000}{10 \times 70}$ = **390 J/(kg °C)**.

Problem 9 5.7 MJ of heat energy are supplied to 30 kg of aluminium which is initially at a temperature of 20°C. If the specific heat capacity of aluminium is 950 J/(kg °C), determine its final temperature.

Quantity of heat, $Q = mc(t_2 - t_1)$
Hence, 5.7 × 10⁶ J = 30 kg × 950 J/(kg °C) × $(t_2 - 20)$°C
from which, $(t_2 - 20) = \dfrac{5.7 \times 10^6}{30 \times 950}$ = 200.

Hence the final temperature, t_2 = 200 + 20 = **220°C**.

Problem 10 Steam initially at a temperature of 130°C is cooled to a temperature of 20°C below the freezing point of water, the loss of heat energy being at a constant rate. Make a sketch, and briefly explain, the expected temperature-time graph representing this change.

A temperature-time graph representing the change is shown in *Fig 5*. Initially steam cools until it reaches the boiling point of water at 100°C. Temperature then remains constant, i.e. between A and B, even though it is still giving off heat

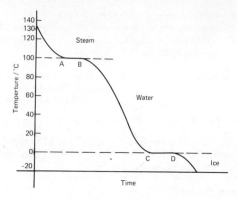

Fig 5

(i.e., latent heat). When all the steam at 100°C has changed to water at 100°C it starts to cool again until it reaches the freezing point of water at 0°C. From C to D the temperature again remains constant until all the water is converted to ice. The temperature of the ice then decreases as shown.

Problem 11 How much heat is needed to completely melt 12 kg of ice at 0°C? Assume the latent heat of fusion of ice is 335 kJ/kg.

Quantity of heat required, $Q = mL$, from para. 6.
$= 12 \text{ kg} \times 335 \text{ kJ/kg}$
$= \mathbf{4020 \text{ kJ} \text{ or } 4.02 \text{ MJ}}.$

Problem 12 Calculate the heat required to convert 5 kg of water at 100°C to superheated steam at 100°C. Assume the latent heat of vaporisation of water is 2260 kJ/kg.

Quantity of heat required, $Q = mL$
$= 5 \text{ kg} \times 2260 \text{ kJ/kg}$
$= \mathbf{11\,300 \text{ kJ} \text{ or } 11.3 \text{ MJ}}.$

Problem 13 Determine the heat energy needed to convert 5 kg of ice initially at −20°C completely to water at 0°C. Assume the specific heat capacity of ice is 2100 J/(kg °C) and the specific latent heat of fusion of ice is 335 kJ/kg.

Quantity of heat energy needed, Q = sensible heat + latent heat.
The quantity of heat needed to raise the temperature of ice from −20°C to 0°C,
i.e. sensible heat, $Q_1 = mc\,(t_2 - t_1) = 5 \text{ kg} \times 2100 \text{ J/(kg °C)} \times (0 - -20)\text{°C}$
$= (5 \times 2100 \times 20) \text{ J} = 210 \text{ kJ}.$
The quantity of heat needed to melt 5 kg of ice at 0°C,
i.e. the latent heat, $Q_2 = mL = 5 \text{ kg} \times 335 \text{ kJ/kg} = 1675 \text{ kJ}.$
Total heat energy needed, $Q = Q_1 + Q_2 = 210 + 1675 = \mathbf{1885 \text{ kJ}}.$

Problem 14 Calculate the heat energy required to convert completely 10 kg of water at 50°C into steam at 100°C, given that the specific heat capacity of water is 4200 J/(kg°C) and the specific latent heat of vaporisation of water is 2260 kJ/kg.

Quantity of heat required = sensible heat + latent heat
Sensible heat, $Q_1 = mc(t_2 - t_1)$ = 10 kg × 4200 J/(kg °C) × (100 − 50)°C
= 2100 kJ
Latent heat, $Q_2 = mL$ = 10 kg × 2260 kJ/kg = 22 600 kJ.
Total heat energy required, $Q = Q_1 + Q_2$ = (2100 + 22 600) kJ
= **24 700 kJ or 24.70 MJ**

Problem 15 State briefly the principle of operation of a refrigerator.

The boiling point of most liquids may be lowered if the pressure is lowered.
In a simple refrigerator a working fluid, such as ammonia or freon, has the pressure acting on it reduced. The resulting lowering of the boiling point causes the liquid to vaporise.

In vaporising, the liquid takes in the necessary latent heat from its surroundings, i.e. the freezer, which thus becomes cooled. The vapour is immediately removed by a pump to a condenser which is outside of the cabinet, where it is compressed and changed back into a liquid, giving out latent heat. The cycle is repeated when the liquid is pumped back to the freezer to be vaporised.

Problem 16 Define the term conduction and give two practical applications.

Conduction is the transfer of heat energy from one part of a body to another, (or from one body to another) without the particles of the body moving. Conduction is associated with solids. For example, if one end of a metal bar is heated, the other end will become hot by conduction.

Metals and metallic alloys are good conductors of heat whereas air, wood, plastic, cork, glass and gases are examples of poor conductors (i.e. heat insulators). Practical applications of conduction:
(i) A domestic saucepan or dish conducts heat from the source to the contents. Also, since wood and plastic are poor conductors of heat they are used for saucepan handles.
(ii) The metal of a radiator of a central heating system conducts heat from the hot water inside to the air outside.

Problem 17 What is meant by convection? Give six examples of convection.

Convection is the transfer of heat energy through a substance by the actual movement of the substance itself. Convection occurs in liquids and gases, but not in solids. When heated, a liquid or gas becomes less dense. It then rises and is replaced by a colder liquid or gas and the process repeats. For example, electric kettles and central heating radiators always heat up at the top first.

Examples of convection are:

(i) Natural circulation hot water heating systems depend on the hot water rising by convection to the top of a house and then falling back to the bottom of the house as it cools, releasing the heat energy to warm the house.
(ii) Convection currents cause air to move and therefore affect climate.
(iii) When a radiator heats the air around it, the hot air rises by convection and cold air moves in to take its place.
(iv) A cooling system in a car radiator relies on convection.
(v) Large electrical transformers dissipate waste heat to an oil tank. The heated oil

rises by convection to the top, then sinks through cooling fins, losing heat as it does so.
(vi) In a refrigerator, the cooling unit is situated near the top. The air surrounding the cold pipes becomes heavier as it contracts and sinks towards the bottom. Warmer, less dense air is pushed upwards and in turn is cooled. A cold convection current is thus created.

Problem 18 Define 'radiation' and state five practical applications of it.

Radiation is the transfer of heat energy from a hot body to a cooler one by electromagnetic waves. Heat radiation is similar in character to light waves (see chapter 11)—it travels at the same speed and can pass through a vacuum—except that the frequency of the waves are different. Waves are emitted by a hot body, are transmitted through space (even a vacuum), and are not detected until they fall on to another body. Radiation is reflected from shining, polished surfaces but absorbed by dull, black surfaces.

Practical applications of radiation include:
(i) heat from the sun reaching earth;
(ii) heat felt by a flame;
(iii) cooker grills;
(iv) industrial furnaces;
(v) infra-red space heaters.

Problem 19 How is a vacuum flask able to keep hot liquids hot and cold liquids cold?

A cross-section of a typical vacuum flask is shown in *Fig 6* and is seen to be a double-walled bottle with a vacuum space between them, the whole supported in a protective outer case.

Very little heat can be transferred by conduction because of the vacuum space and the cork stopper (cork is a bad conductor of heat). Also, because of the vacuum space, no convection is possible. Radiation is minimised by silvering the two glass surfaces (radiation is reflected off of shining surfaces).

Thus a vacuum flask is an example of prevention of all three types of heat transfer and is therefore able to keep hot liquids hot and cold liquids cold.

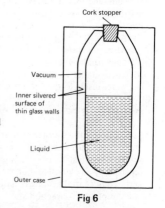

Fig 6

Problem 20 Give four examples of how insulation may be used to conserve fuel in a domestic heating installation.

Fuel used for heating a building is becoming increasingly expensive. By the careful use of insulation, heat can be retained in a building for longer periods and the cost of heating thus minimised.
(i) Since convection causes hot air to rise it is important to insulate the roof space,

which is probably the greatest source of heat loss in the home. This can be achieved by laying fibre-glass between the wooden joists in the roof space.
(ii) Glass is a poor conductor of heat. However large losses can occur through thin panes of glass and such losses can be reduced by using double-glazing. Two sheets of glass, separated by air, are used. Air is a good insulator but the air space must not be too large otherwise convection currents can occur which would carry heat across the space.
(iii) Hot water tanks should be lagged to prevent conduction and convection of heat to the surrounding air.
(iv) Brick, concrete, plaster and wood are all poor conductors of heat. A house is made from two walls with an air gap between them. Air is a poor conductor and trapped air minimises losses through the walls. Heat losses through walls can be prevented almost completely by using cavity wall insulation, i.e. plastic-foam.

Problem 21 Give examples of practical applications, and design implications, of thermal movement.

Most substances expand when heated and contract when cooled, each effect depending on the change of temperature of the substance.
(i) Overhead electrical transmission lines are hung so that they are slack in summer, otherwise their contraction in winter might snap the conductors or bring down pylons.
(ii) Gaps need to be left in lengths of railway lines to prevent buckling in hot weather.
(iii) Ends of large bridges are often supported on rollers to allow them to expand and contract freely.
(iv) Fitting a metal collar to a shaft or a steel tyre to a wheel is often achieved by first heating them so that they expand, fitting them in position, and then cooling them so that the contraction holds them firmly in place. This is known as a 'shrink-fit'. By a similar method hot rivets are used for joining metal sheets.
(v) Thermometers use the expansion of a liquid such as mercury or alcohol to measure temperature.
(vi) The amount of expansion varies with different materials. *Fig 7(a)* shows a bimetallic strip at room temperature, (i.e., two different strips of metal riveted together).

When heated, brass expands more than steel, and since the two metals are riveted together the bimetallic strip is forced into an arc as shown in *Fig 7(b)*. Such a movement can be arranged to make or break an electric circuit and bimetallic strips are used, in particular, in thermostats (which are temperature operated switches) used to control central heating systems, cookers, refrigerators, toasters, irons, hot-water and alarm systems.

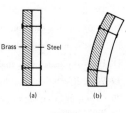

Fig 7

(vii) Motor engines use the rapid expansion of heated gases to force a piston to move.
(viii) Water is a liquid which at low temperatures displays an unusual effect. If cooled, contraction occurs until, at about 4°C, the volume is at a minimum. As the temperature is further decreased from 4°C to 0°C expansion occurs, i.e.,

the volume increases. When ice is formed considerable expansion occurs and it is this expansion which often causes frozen water pipes to burst.

(ix) Designers must predict, and allow for, the expansion of steel pipes in a steam-raising plant so as to avoid damage and consequent danger to health.

C. FURTHER PROBLEMS ON HEAT ENERGY

(a) SHORT ANSWER PROBLEMS

1 Differentiate between temperature and heat.

2 Name two scales on which temperature is measured.

3 How are the fixed points on the Celsius scale obtained?

4 Briefly describe the principle of operation of a mercury-in-glass thermometer.

5 Make a comparison between mercury and alcohol as the liquid in a liquid-in-glass thermometer.

6 What is a thermocouple?

7 Briefly describe how a thermocouple can be used with a galvanometer for temperature measurement.

8 List three advantages of a thermocouple compared with other temperature measuring devices.

9 Define specific heat capacity and name its unit.

10 Differentiate between sensible and latent heat.

11 The quantity of heat, Q, required to raise a mass m kg from temperature $t_1\,°C$ to $t_2\,°C$, the specific heat capacity being c, is given by $Q = \ldots\ldots\ldots\ldots$.

12 What is meant by the specific latent heat of fusion?

13 Define the specific latent heat of vaporisation.

14 State three methods of heat transfer.

15 Define conduction and state two practical examples of heat transfer by this method.

16 Define convection and give three examples of heat transfer by this method.

17 What is meant by radiation? Give three uses.

18 How can insulation conserve fuel in a typical house?

19 What effects does heat have on the physical dimensions of solids, liquids and gases?

20 Give four examples of practical applications and design implications of thermal movement.

(b) MULTI-CHOICE PROBLEMS (answers on page 149)

1 Heat energy is measured in:
(a) kelvin; (b) watts; (c) kilograms; (d) joules.

2 A change of temperature of 20°C is equivalent to a change of thermodynamic temperature of (a) 293 K; (b) 20 K.

3 A temperature of 20°C is equivalent to (a) 293 K; (b) 20 K.

4 The unit of specific heat capacity is:
   (a) joules per kilogram;   (c) joules per kilogram kelvin;
   (b) joules;   (d) cubic metres.

5 Which of the following statements is false?
   (a) −30°C is equivalent to 243 K.
   (b) Convection only occurs in liquids and gases.
   (c) Conduction and convection cannot occur in a vacuum.
   (d) Radiation is absorbed by a silver surface.

6 The quantity of heat required to raise the temperature of 500 g of iron by 2°C, given that the specific heat capacity is 500 J/(kg °C), is:
   (a) 500 kJ; (b) 0.5 kJ; (c) 2 J; (d) 250 kJ.

7 The heat energy required to change 1 kg of a substance from a liquid to a gaseous state at the same temperature is called:
   (a) specific heat capacity, (b) specific latent heat of vaporisation,
   (c) sensible heat, (d) specific latent heat of fusion.

8 When two wires of different materials are joined together and heat applied to the junction an emf is produced. This effect is used in a thermocouple to measure:
   (a) heat; (b) emf; (c) temperature; (d) expansion.

9 The transfer of heat energy through a substance by the actual movement of the particles of the substance is called:
   (a) conduction; (b) radiation; (c) convection; (d) specific heat capacity.

10 Which of the following statements is true?
   (a) Heat is the degree of hotness or coldness of a body.
   (b) Heat energy which flows to or from a substance while the temperature remains constant is called sensible heat.
   (c) The unit of the specific latent heat of fusion is J/(kg K).
   (d) A cooker-grill is a practical application of radiation.

(c) CONVENTIONAL PROBLEMS

1 Convert the following temperatures into the kelvin scale:
   (a) 51°C; (b) −78°C; (c) 183°C.   [(a) 324 K; (b) 195 K; (c) 456 K]

2 Convert the following temperatures into the Celsius scale:
   (a) 307 K; (b) 237 K; (c) 415 K.   [(a) 34°C; (b) −36°C; (c) 142°C]

3 (a) What is the difference between heat and temperature?
   (b) State three temperature measuring devices and state the principle of operation of each.

4 Determine the quantity of heat energy (in megajoules) required to raise the temperature of 10 kg of water from 0°C to 50°C. Assume the specific heat capacity of water is 4200 J/(kg °C).   [2.1 MJ]

5 Some copper, having a mass of 20 kg, cools from a temperature of 120°C to 70°C. If the specific heat capacity of copper is 390 J/(kg °C), how much heat energy is lost by the copper?   [390 kJ]

6 A block of aluminium having a specific heat capacity of 950 J/(kg °C) is heated

from 60°C to its melting point at 660°C. If the quantity of heat required is 2.85 MJ, determine the mass of the aluminium block. [5 kg]

7 20.8 kJ of heat energy is required to raise the temperature of 2 kg of lead from 16°C to 96°C. Determine the specific heat capacity of lead. [130 J/(kg °C)]

8 250 kJ of heat energy is supplied to 10 kg of iron which is initially at a temperature of 15°C. If the specific heat capacity of iron is 500 J/(kg °C), determine its final temperature. [65°C]

9 Some ice, initially at −40°C has heat supplied to it at a constant rate until it becomes superheated steam at 150°C. Sketch a typical temperature-time graph expected and use it to explain the difference between sensible and latent heat.

10 How much heat is needed to completely melt 25 kg of ice at 0°C. Assume the specific latent heat of fusion of ice is 335 kJ/kg. [8.375 MJ]

11 Determine the heat energy required to change 8 kg of water at 100°C to superheated steam at 100°C. Assume the specific latent heat of vaporisation of water is 2260 kJ/kg. [18.08 MJ]

12 Calculate the heat energy required to convert 10 kg of ice initially at −30°C completely into water at 0°C. Assume the specific heat capacity of ice is 2.1 kJ/(kg °C) and the specific latent heat of fusion of ice is 335 kJ/kg.

[3.98 MJ]

13 Determine the heat energy needed to convert completely 5 kg of water at 60°C to steam at 100°C, given that the specific heat capacity of water is 4.2 kJ/(kg °C) and the specific latent heat of vaporisation of water is 2260 kJ/kg. [12.14 MJ]

14 Describe three methods of heat transfer and state one practical example of each.

15 State some possible precautions in preventing heat loss from a building.

16 State five practical applications and design implications of thermal movement.

14 Simple electric circuits

A. MAIN POINTS CONCERNED WITH SIMPLE ELECTRIC CIRCUITS

1 **Standard symbols for electrical components**
   Symbols are used for components in electrical circuit diagrams and some of the more common ones are shown in *Fig 1*.
2 (i) All substances are made from **elements** and the smallest particle to which an element can be reduced is called an **atom**.
   (ii) An atom consists of **electrons** which can be considered to be orbiting around a central **nucleus** containing **protons** and **neutrons**.
   (iii) An electron possesses a **negative charge**, a proton **a positive charge** and a neutron has **no charge**.

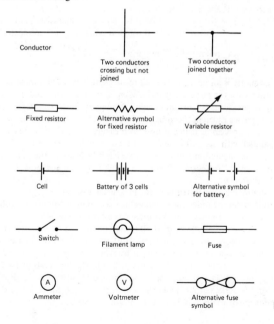

Fig 1

(iv) There is a force of **attraction** between oppositely charged bodies and a force of **repulsion** between similarly charged bodies.

(v) The **force** between two charged bodies depends on the amount of charge on the bodies and their distance apart.

(vi) **Conductors** are materials that have electrons that are loosely connected to the nucleus and can easily move through the material from one atom to another. **Insulators** are materials whose electrons are held firmly to their nucleus.

(vii) A drift of electrons in the same direction constitutes an **electric current**.

(viii) The unit of charge is the **coulomb, C**, and when 1 coulomb of charge is transferred in 1 second a current of 1 ampere flows in the conductor.

Thus electric current I is the rate of flow of charge in a circuit. The unit of current is the **ampere, A**.

(ix) For a continuous current to flow between two points in a circuit a **potential difference (p.d.)** or **voltage, V**, is required between them; a complete conducting path is necessary to and from the source of electrical energy. The unit of p.d. is the **volt, V**.

(x) *Fig 2* shows a cell connected across a filament lamp. Current flow, by convention, is considered as flowing from the positive terminal of the cell, around the circuit to the negative terminal.

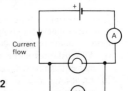

Fig 2

3 The flow of electric current is subject to friction. This friction, or opposition, is called **resistance R** and is the property of a conductor that limits current. The unit unit of resistance is the **ohm, Ω**. 1 ohm is defined as the resistance which will have a current of 1 ampere flowing through it when 1 volt is connected across it,

i.e. resistance $R = \dfrac{\text{potential difference}}{\text{current}}$.

4 (i) An **ammeter** is an instrument used to measure current and must be connected **in series** with the circuit. *Fig 2* shows an ammeter connected in series with the lamp to measure the current flowing through it. Since all the current in the circuit passes through the ammeter it must have a very **low resistance**.

(ii) A voltmeter is an instrument used to measure p.d. and must be connected **in parallel** with the part of the circuit whose p.d. is required. In *Fig 2*, a voltmeter is connected in parallel with the lamp to measure the p.d. across it. To avoid a significant current flowing through it a voltmeter must have a very **high resistance**.

(iii) An **ohmmeter** is an instrument for measuring resistance.

(iv) A **multimeter**, or universal instrument, may be used to measure voltage, current and resistance. An 'Avometer' is a typical example.

(v) The **cathode ray oscilloscope (CRO)** may be used to observe waveforms and to measure voltages and currents. The display of a CRO involves a spot of light moving across a screen. The amount by which the spot is deflected from its initial position depends on the p.d. applied to the terminals of the CRO and the range selected. The displacement is calibrated in 'volts per cm'. For example, if the spot is deflected 3 cm and the volts/cm switch is on 10 V/cm then the magnitude of the p.d. is 3 cm × 10 V/cm, i.e. 30 V.

5 *Fig 3* shows a circuit in which current I can be varied by the variable resistor R_2. For various settings of R_2, the current flowing in resistor R_1, displayed on the ammeter, and the p.d. across R_1, displayed on the voltmeter, are noted and a graph is plotted of p.d. against current. The result is shown in *Fig 4(a)* where the straight line graph passing through the origin indicates that current is directly proportional to the p.d. Since the gradient

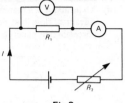

Fig 3

i.e. $\dfrac{\text{p.d.}}{\text{current}}$

is constant, resistance R_1 is constant. A resistor is thus an example of a **linear device**.

If the resistor R_1 in *Fig 3* is replaced by a component such as a lamp then the graph shown in *Fig 4(b)* results when values of p.d. are noted for various current readings. Since the gradient is changing the lamp is an example of a **non-linear device**.

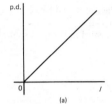
(a)

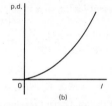
(b)

Fig 4

6 **Ohm's law** states that the current I flowing in a circuit is directly proportional to the applied voltage V and inversely proportional to the resistance R, provided the resistance remains constant. Thus,

$$I = \frac{V}{R} \text{ or } V = IR \text{ or } R = \frac{V}{I}.$$

7 Currents, voltages and resistances can often be very large or very small. Thus **multiples and submultiples** of units are often used. The most common ones are listed in *Table 1*.

TABLE 1

Prefix	Name	Meaning	Example
M	mega	multiply by 1 000 000 (i.e., $\times 10^6$)	2 MΩ = 2 000 000 ohms
k	kilo	multiply by 1000 (i.e., $\times 10^3$)	10 kV = 10 000 volts
m	milli	divide by 1000 (i.e., $\times 10^{-3}$)	25 mA = $\dfrac{25}{1000}$ A = 0.025 amperes
μ	micro	divide by 1 000 000 (i.e., $\times 10^{-6}$)	50 μV = $\dfrac{50}{1\,000\,000}$ V = 0.00005 volts

8. (i) A **conductor** is a material having a low resistance which allows electric current to flow in it. All metals are conductors and some examples include copper, aluminium, brass, platinum, silver, gold and also carbon.
   (ii) An **insulator** is a material having a high resistance which does not allow electric current to flow in it. Some examples of insulators include plastic, rubber, glass, porcelain, air, paper, cork, mica, ceramics and certain oils.

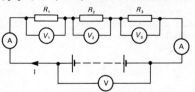

Fig 5

9. **Series circuit**

Fig 5 shows three resistors R_1, R_2 and R_3 connected end to end, i.e., in series, with a battery source of V volts. Since the circuit is closed a current I will flow and the p.d. across each resistor may be determined from the voltmeter readings V_1, V_2 and V_3.

In a series circuit:
(a) the current I is the same in all parts of the circuit and hence the same reading is found on each of the ammeters shown, and
(b) the sum of the voltages V_1, V_2 and V_3 is equal to the total applied voltage, V,

i.e. $V = V_1 + V_2 + V_3$.

From Ohm's law:

$V_1 = IR_1$, $V_2 = IR_2$, $V_3 = IR_3$ and $V = IR$

where R is the total circuit resistance.
Since $V = V_1 + V_2 + V_3$
then $IR = IR_1 + IR_2 + IR_3$
Dividing throughout by I gives
$R = R_1 + R_2 + R_3$.

Thus for a series circuit, the total resistance is obtained by adding together the values of the separate resistances.

10. **Parallel circuit**

Fig 6 shows three resistors, R_1, R_2 and R_3 connected across each other, i.e., in parallel, across a battery source of V volts. In a parallel circuit:
(a) the sum of the currents I_1, I_2 and I_3 is equal to the total circuit current, I,
i.e. $I = I_1 + I_2 + I_3$, and
(b) the source p.d., V volts, is the same across each of the resistors.

From Ohm's law:

$I_1 = \dfrac{V}{R_1}$, $I_2 = \dfrac{V}{R_2}$, $I_3 = \dfrac{V}{R_3}$ and $I = \dfrac{V}{R}$

where R is the total circuit resistance.
Since $I = I_1 + I_2 + I_3$

Then, $\dfrac{V}{R} = \dfrac{V}{R_1} + \dfrac{V}{R_2} + \dfrac{V}{R_3}$

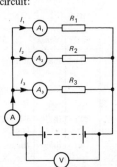

Fig 6

Dividing throughout by V gives:

$$\frac{1}{R} = \frac{1}{R_1} + \frac{1}{R_2} + \frac{1}{R_3}$$

This equation must be used when finding the total resistance R of a parallel circuit. For the special case of two resistors in parallel

$$\frac{1}{R} = \frac{1}{R_1} + \frac{1}{R_2} = \frac{R_2 + R_1}{R_1 R_2}$$

Hence $R = \dfrac{R_1 R_2}{R_1 + R_2}$ [i.e. $\dfrac{\text{product}}{\text{sum}}$]

11 Wiring lamps in series and in parallel

Series connection

Fig 7 shows three lamps, each rated at 240 V, connected in series across a 240 V supply.

(i) Each lamp has only $\dfrac{240}{3}$ V, i.e., 80 V across it and thus each lamp glows dimly.

(ii) If another lamp of similar rating is added in series with the other three lamps then each lamp now has $\dfrac{240}{4}$ V, i.e. 60 V across it and each now glows even more dimly.

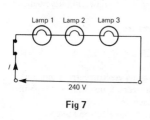

Fig 7

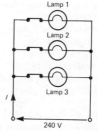

Fig 8

(iii) If a lamp is removed from the circuit or if a lamp develops a fault (i.e. an open circuit) or if the switch is opened then the circuit is broken, no current flows, and the remaining lamps will not light up.

(iv) Less cable is required for a series connection than for a parallel one.

The series connection of lamps is usually limited to decorative lighting such as for Christmas tree lights.

Parallel connection

Fig 8 shows three similar lamps, each rated at 240 V, connected in parallel across a 240 V supply.

(i) Each lamp has 240 V across it and thus each will glow brilliantly at their rated voltage.

(ii) If any lamp is removed from the circuit or develops a fault (open circuit) or a switch is opened, the remaining lamps are unaffected.

(iii) The addition of further similar lamps in parallel does not affect the brightness of the other lamps.

(iv) More cable is required for parallel connection than for a series one.

The parallel connection of lamps is the most widely used in electrical installations.

12 **Power P** in an electrical circuit is given by the product of potential difference V and current I. The unit of power is the **watt, W**.

Hence $P = V \times I$ watts   (1)

From Ohm's law, $V = IR$
Substituting for V in (1) gives:

$P = (IR) \times I$

i.e. $P = I^2 R$ **watts**

Also, from Ohm's law, $I = \dfrac{V}{R}$

Substituting for I in (1) gives:

$P = V \times \left(\dfrac{V}{R}\right)$

i.e. $P = \dfrac{V^2}{R}$ watts.

There are thus three possible formulae which may be used for calculating power.

13 **Electrical energy** = power × time.

If the power is measured in watts and the time in seconds then the unit of energy is watt-seconds or **joules**. If the power is measured in kilowatts and the time in hours then the unit of energy is **Kilowatt-hours**, often called the **'unit of electricity'**. The 'electricity meter' in the home records the number of kilowatt-hours used and is thus an energy meter.

14 (i) The three main effects of an electric current are:
    (a) magnetic effect;
    (b) chemical effect;
    (c) heating effect.

  (ii) Some practical applications of the effects of an electric current include:
    Magnetic effect: bells, relays, motors, generators, transformers, telephones, car-ignition and lifting magnets.
    Chemical effect: primary and secondary cells and electroplating.
    Heating effect: cookers, water heaters, electric fires, irons, furnaces, kettles and soldering irons.

15 A **fuse** is used to prevent overloading of electrical circuits. The fuse, which is made of material having a low melting point, utilizes the heating effect of an electric current. A fuse is placed in an electrical circuit and if the current becomes too large the fuse wire melts and so breaks the circuit. A circuit diagram symbol for a fuse is shown in *Fig 1*.

B. WORKED PROBLEMS ON SIMPLE ELECTRIC CIRCUITS

Problem 1 The current flowing through a resistor is 0.8 A when a p.d. of 20 V is applied. Determine the value of the resistance.

From Ohm's law, resistance $R = \dfrac{V}{I} = \dfrac{20}{0.8} = \dfrac{200}{8} = 25\ \Omega$.

Problem 2 Determine the p.d. which must be applied to a 2 kΩ resistor in order that a current of 10 mA may flow.

Resistance $R = 2\ \text{k}\Omega = 2 \times 10^3\ \Omega = 2000\ \Omega$

Current $I = 10\ \text{mA} = 10 \times 10^{-3}\ \text{A}$ or $\dfrac{10}{10^3}\ \text{A}$ or $\dfrac{10}{1000}\ \text{A} = 0.01\ \text{A}$

From Ohm's law, potential difference, $V = IR = (0.01)(2000) =$ **20 V**

Problem 3 The hot resistance of a 240 V filament lamp is 960 Ω. Find the current taken by the lamp and its power rating.

From Ohm's law, current $I = \dfrac{V}{R} = \dfrac{240}{960} = \dfrac{24}{96} = \dfrac{1}{4}$ **A or 0.25 A**

Power rating $P = VI = (240)(\tfrac{1}{4}) =$ **60 W**

Problem 4 What is the resistance of a coil which draws a current of (a) 50 mA and (b) 200 μA from a 120 V supply?

(a) Resistance $R = \dfrac{V}{I} = \dfrac{120}{50 \times 10^{-3}}$

$= \dfrac{120}{0.05} = \dfrac{12\,000}{5} =$ **2400 Ω or 2.4 kΩ**

(b) Resistance $R = \dfrac{120}{200 \times 10^{-6}} = \dfrac{120}{0.0002}$

$= \dfrac{1\,200\,000}{2} =$ **600 000 Ω or 600 kΩ or 0.6 MΩ**

Problem 5 A 12 V battery is connected across a load having a resistance of 40 Ω. Determine the current flowing in the load, the power consumed and the energy dissipated in 2 minutes.

Current $I = \dfrac{V}{R} = \dfrac{12}{40} =$ **0.3 A**

Power consumed, $P = VI = (12)(0.3) =$ **3.6 W**

Energy dissipated = power $\times$ time = $(3.6\ \text{W})(2 \times 60\ \text{s}) =$ **432 J** (since 1 J = 1 W s)

Problem 6 A 100 V battery is connected across a resistor and causes a current of 5 mA to flow. Determine the resistance of the resistor. If the voltage is now reduced to 25 V, what will be the new value of the current flowing?

Resistance $R = \dfrac{V}{I} = \dfrac{100}{5 \times 10^{-3}} = \dfrac{100 \times 10^3}{5} = 20 \times 10^3 =$ **20 kΩ**

Current when voltage is reduced to 25 V,

$I = \dfrac{V}{R} = \dfrac{25}{20 \times 10^3} = \dfrac{25}{20} \times 10^{-3} =$ **1.25 mA**

Problem 7 For the circuit shown in *Fig 9*, determine (a) the battery voltage V, (b) the total resistance of the circuit and (c) the values of resistance of resistors R_1, R_2 and R_3, given that the p.d.'s across R_1, R_2 and R_3 are 5 V, 2 V and 6 V respectively.

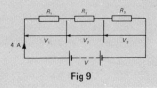

Fig 9

(a) Battery voltage $V = V_1 + V_2 + V_3$
$= 5 + 2 + 6 = $ **13 V**

(b) Total circuit resistance $R = \dfrac{V}{I} = \dfrac{13}{4} = $ **3.25 Ω**

(c) Resistance $R_1 = \dfrac{V_1}{I} = \dfrac{5}{4} = $ **1.25 Ω**

Resistance $R_2 = \dfrac{V_2}{I} = \dfrac{2}{4} = $ **0.5 Ω**

Resistance $R_3 = \dfrac{V_3}{I} = \dfrac{6}{4} = $ **1.5 Ω**

(*Check*: $R_1 + R_2 + R_3 = 1.25 + 0.5 + 1.5 = 3.25 \, \Omega = R$)

Problem 8 For the circuit shown in *Fig 10* determine the p.d. across resistor R_3. If the total resistance of the circuit is 100 Ω, determine the current flowing through resistor R_1. Find also the value of resistor R_2.

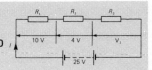

Fig 10

P.d. across R_3, $V_3 = 25 - 10 - 4 = $ **11 V**

Current $I = \dfrac{V}{R} = \dfrac{25}{100} = $ **0.25 A**, which is the current flowing in each resistor.

Resistance $R_2 = \dfrac{V_2}{I} = \dfrac{4}{0.25} = $ **16 Ω**

Problem 9 The current/voltage relationship for two resistors A and B is as shown in *Fig 11*. Determine the value of the resistance of each resistor.

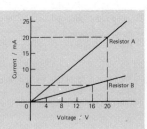

Fig 11

For resistor A, $R = \dfrac{V}{I} = \dfrac{20 \text{ V}}{20 \text{ mA}} = \dfrac{20}{0.02} = \dfrac{2000}{2} = $ **1000 Ω or 1 kΩ**

For resistor B, $R = \dfrac{V}{I} = \dfrac{16 \text{ V}}{5 \text{ mA}} = \dfrac{16}{0.005} = \dfrac{16\,000}{5} = $ **3200 Ω or 3.2 kΩ**

Problem 10 A 12 V battery is connected in a circuit having three series-connected resistors having resistance of 4 Ω, 9 Ω and 11 Ω. Determine the current flowing through, and the p.d. across the 9 Ω resistor. Find also the power dissipated in the 11 Ω resistor.

The circuit diagram is shown in *Fig 12*.
Total resistance $R = 4 + 9 + 11 = 24$ Ω

Current $I = \dfrac{V}{R} = \dfrac{12}{24} = $ **0.5 A**,

which is the current in the 9 Ω resistor.
P.d. across the 9 Ω resistor,
$V_1 = I \times 9 = 0.5 \times 9 = $ **4.5 V**

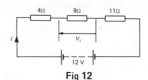

Fig 12

Power dissipated in the 11 Ω resistor, $P = I^2 R = 0.5^2(11) = (0.25)(11) = $ **2.75 W**

Problem 11 Two resistors are connected in series across a 24 V supply and a current of 3 A flows in the circuit. If one of the resistors has a resistance of 2 Ω determine (a) the value of the other resistor, and (b) the p.d. across the 2 Ω resistor. If the circuit is connected for 50 hours, how much energy is used?

The circuit diagram is shown in *Fig 13*.

(a) Total circuit resistance

$R = \dfrac{V}{I} = \dfrac{24}{3} = 8$ Ω

Value of unknown resistance
$R_x = 8 - 2 = $ **6 Ω**

(b) P.d. across 2 Ω resistor,
$V_1 = IR_1 = 3 \times 2 = $ **6 V**
Energy used = power × time
= $V \times I \times t$
= $(24 \times 3 \text{ W})(50 \text{ h})$
= 3600 Wh = **3.6 kW h**

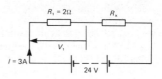

Fig 13

Problem 12 For the circuit shown in *Fig 14*, determine (a) the reading on the ammeter, and (b) the value of resistor R_2.

Fig 14

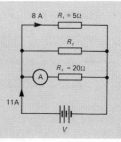

P.d. across R_1 is the same as the supply voltage V.
Hence supply voltage, $V = 8 \times 5 = 40$ V

(a) Reading on ammeter, $I = \dfrac{V}{R_3} = \dfrac{40}{20} = $ **2 A**

(b) Current flowing through $R_2 = 11 - 8 - 2 = 1$ A

Hence, $R_2 = \dfrac{V}{I_2} = \dfrac{40}{1} =$ **40 Ω**

Problem 13 Two resistors, of resistance 3 Ω and 6 Ω, are connected in parallel across a battery having a voltage of 12 V. Determine (a) the total circuit resistance and (b) the current flowing in the 3 Ω resistor.

The circuit diagram is shown in *Fig 15*.

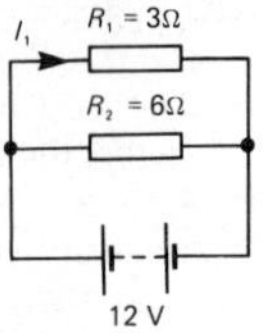

(a) The total circuit resistance R is given by

$$\dfrac{1}{R} = \dfrac{1}{R_1} + \dfrac{1}{R_2} = \dfrac{1}{3} + \dfrac{1}{6}$$

$$\dfrac{1}{R} = \dfrac{2+1}{6} = \dfrac{3}{6}$$

Hence, $R = \dfrac{6}{3} =$ **2 Ω**

Fig 15

[Alternatively, $R = \dfrac{R_1 R_2}{R_1 + R_2} = \dfrac{3 \times 6}{3 + 6} = \dfrac{18}{9} = 2$ Ω (see para. 10).]

(b) Current in the 3 Ω resistance, $I_1 = \dfrac{V}{R_1} = \dfrac{12}{3} =$ **4 A**

Problem 14 For the circuit shown in *Fig 16*, find (a) the value of the supply voltage V and (b) the value of current I.

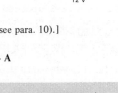

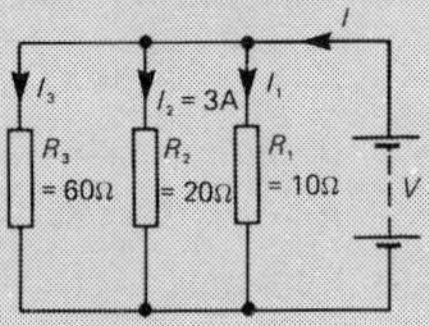

Fig 16

(a) P.d. across 20 Ω resistor $= I_2 R_2 = 3 \times 20 = 60$ V.
Hence supply voltage $V =$ **60 V** since the circuit is connected in parallel.

(b) Current $I_1 = \dfrac{V}{R_1} = \dfrac{60}{10} = 6$ A; $I_2 = 3$ A; $I_3 = \dfrac{V}{R_3} = \dfrac{60}{60} = 1$ A

Current $I = I_1 + I_2 + I_3$ and hence $I = 6 + 3 + 1 =$ **10 A**

Alternatively, $\dfrac{1}{R} = \dfrac{1}{60} + \dfrac{1}{20} + \dfrac{1}{10} = \dfrac{1+3+6}{60} = \dfrac{10}{60}$

Hence total resistance $R = \dfrac{60}{10} = 6$ Ω

Current $I = \dfrac{V}{R} = \dfrac{60}{6} =$ **10 A**

Problem 15 If three identical lamps are connected in parallel and the combined resistance is 150 Ω, find the resistance of one lamp.

Let the resistance of one lamp be R,

Then, $\dfrac{1}{150} = \dfrac{1}{R} + \dfrac{1}{R} + \dfrac{1}{R} = \dfrac{3}{R}$

from which, $R = 3 \times 150 = 450$ Ω

Problem 16 Given four 1 Ω resistors, state how they must be connected to give an overall resistance of (a) $\frac{1}{4}$Ω; (b) 1 Ω; (c) $1\frac{1}{3}$Ω; (d) $2\frac{1}{2}$Ω, all four resistors being connected in each case.

(a) **All four in parallel** (see *Fig 17*),

since $\frac{1}{R} = \frac{1}{1} + \frac{1}{1} + \frac{1}{1} + \frac{1}{1} = \frac{4}{1}$, i.e., $R = \frac{1}{4}$Ω

(b) **Two in series, in parallel with another two in series** (see *Fig 18*), since 1 Ω and 1 Ω in series gives 2 Ω, and 2 Ω in parallel with 2 Ω gives

$\frac{2 \times 2}{2 + 2} = \frac{4}{4} = 1$ Ω

(c) **Three in parallel in series with one** (see *Fig 19*), since for the three in parallel,

$\frac{1}{R} = \frac{1}{1} + \frac{1}{1} + \frac{1}{1} = \frac{3}{1}$, i.e., $R = \frac{1}{3}$Ω and $\frac{1}{3}$Ω in series with 1 Ω gives $1\frac{1}{3}$Ω.

(d) **Two in parallel, in series with two in series** (see *Fig 20*), since for the two in parallel

$R = \frac{1 \times 1}{1 + 1} = \frac{1}{2}$Ω, and $\frac{1}{2}$Ω, 1 Ω and 1 Ω in series gives $2\frac{1}{2}$Ω.

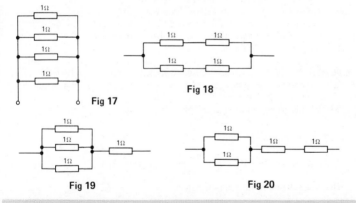

Fig 17 Fig 18 Fig 19 Fig 20

Problem 17 Three identical lamps A, B and C are connected in series across a 150 V supply. State (a) the voltage across each lamp, and (b) the effect of lamp C failing.

(a) Since each lamp is identical and they are connected in series there is $\frac{150}{3}$ V, i.e. **50 V** across each.
(b) If lamp C fails, i.e., open circuits, no current will flow and **lamps A and B will not operate**.

Problem 18 A 100 W electric light bulb is connected to a 250 V supply. Determine (a) the current flowing in the bulb, and (b) the resistance of the bulb.

Power $P = V \times I$, from which, current $I = \dfrac{P}{V}$

(a) Current $I = \dfrac{100}{250} = \dfrac{10}{25} = \dfrac{2}{5} = \textbf{0.4 A}$

(b) Resistance $R = \dfrac{V}{I} = \dfrac{250}{0.4} = \dfrac{2500}{4} = \textbf{625} \, \boldsymbol{\Omega}$

Problem 19 Calculate the power dissipated when a current of 4 mA flows through a resistance of 5 kΩ.

Power $P = I^2 R = (4 \times 10^{-3})^2 (5 \times 10^3)$
$= 16 \times 10^{-6} \times 5 \times 10^3 = 80 \times 10^{-3}$
$= \textbf{0.08 W or 80 mW}$

Alternatively, since $I = 4 \times 10^{-3}$ and $R = 5 \times 10^3$ then from Ohm's law
voltage $V = IR = 4 \times 10^{-3} \times 5 \times 10^3 = 20$ V.
Hence, power $P = V \times I = 20 \times 4 \times 10^{-3} = \textbf{80 mW}$.

Problem 20 Determine the power dissipated by the element of an electric fire of resistance 20Ω when a current of 10 A flows through it. If the fire is on for 6 hours determine the energy used and the cost if 1 unit of electricity costs 5p.

Power $P = I^2 R = 10^2 \times 20 = 100 \times 20 = \textbf{2000 W or 2 kW}$.
Alternatively, from Ohm's law, $V = IR = 10 \times 20 = 200$ V
Hence power $P = V \times I = 200 \times 10 = 2000$ W $= \textbf{2 kW}$
Energy used in 6 hours = power × time = 2 kW × 6 h = **12 kWh**.
1 unit of electricity = 1 kWh. Hence the number of units used is 12.
Cost of energy = 12 × 5 = **60p**.

Problem 21 If 5 A, 10 A and 13 A fuses are available, state which is most appropriate for the following appliances which are both connected to a 240 V supply.
(a) Electric toaster having a power rating of 1 kW.
(b) Electric fire having a power rating of 3 kW.

Power $P = VI$, from which, current $I = \dfrac{P}{V}$

(a) For the toaster, current $I = \dfrac{P}{V} = \dfrac{1000}{240} = \dfrac{100}{24} = 4\dfrac{1}{6}$ A

Hence a **5 A fuse** is most appropriate.

(b) For the fire, current $I = \dfrac{P}{V} = \dfrac{3000}{240} = \dfrac{300}{24} = 12\dfrac{1}{2}$ A

Hence a **13 A fuse** is most appropriate.

Problem 22 A current of 5 A flows in the winding of an electric motor, the resistance of the winding being 100 Ω. Determine (a) the p.d. across the winding, and (b) the power dissipated by the coil.

(a) Potential difference across winding, $V = IR = 5 \times 100 = \textbf{500 V}$
(b) Power dissipated by coil, $P = I^2 R = 5^2 \times 100 = \textbf{2500 W or 2.5 kW}$
 (Alternatively, $P = V \times I = 500 \times 5 = \textbf{2500 W or 2.5 kW}$)

Problem 23 A business uses two 3 kW fires for an average of 20 hours each per week, and six 150 W lights for 30 hours each per week. If the cost of electricity is 5p per unit, determine the weekly cost of electricity to the business.

Energy = power × time
Energy used by one 3 kW fire in 20 hours = 3 kW × 20 h = 60 kWh
Hence weekly energy used by two 3 kW fires = 2 × 60 = 120 kWh
Energy used by one 150 W light for 30 hours = 150 W × 30 h = 4500 Wh = 4.5 kWh
Hence weekly energy used by six 150 W lamps = 6 × 4.5 = 27 kWh
Total energy used per week = 120 + 27 = 147 kWh
1 unit of electricity = 1 kWh of energy
Thus weekly cost of energy at 5p per kWh = 5 × 147 = 735p = **£7.35**

C. FURTHER PROBLEMS ON SIMPLE ELECTRIC CIRCUITS

(a) SHORT ANSWER PROBLEMS

1. Draw the preferred symbols for the following components used when drawing electrical circuit diagrams:
   (a) fixed resistor; (b) cell; (c) filament lamp; (d) fuse; (e) voltmeter.

2. State the unit of (a) current, (b) potential difference, (c) resistance.

3. State an instrument used to measure (a) current, (b) potential difference, (c) resistance.

4. What is a multimeter?

5. State Ohm's law.

6. Give one example of (a) a linear device, (b) a non-linear device.

7. State the meaning of the following abbreviations of prefixes used with electrical units: (a) k, (b) μ, (c) m, (d) M.

8. What is a conductor? Give four examples.

9. What is an insulator? Give four examples.

10. Complete the following statement:
    An ammeter has a .... resistance and must be connected .... with the load.

11. Complete the following statement:
    A voltmeter has a .... resistance and must be connected .... with the load.

12. Show that for three resistors R_1, R_2 and R_3 connected in series the equivalent resistance R is given by $R = R_1 + R_2 + R_3$.

13. Show that for three resistors R_1, R_2 and R_3 connected in parallel the equivalent resistance R is given by $\frac{1}{R} = \frac{1}{R_1} + \frac{1}{R_2} + \frac{1}{R_3}$.

14. Compare the merits of wiring lamps in (a) series, (b) parallel.

15 State the unit of electrical power. State three formulae used to calculate power.

16 State two units used for electrical energy.

17 State the three main effects of an electric current and give two examples of each.

18 What is the function of a fuse in an electrical circuit?

(b) MULTI-CHOICE PROBLEMS (answers on page 149)

1 If two 4 Ω resistors are connected in series the effective resistance of the circuit is:
 (a) 8 Ω; (b) 4 Ω; (c) 2 Ω; (d) 1 Ω.

2 If two 4 Ω resistors are connected in parallel the effective resistance of the circuit is:
 (a) 8 Ω; (b) 4 Ω; (c) 2 Ω; (d) 1 Ω.

3 With the switch in *Fig 21* closed, the ammeter reading will indicate:
 (a) $1\frac{2}{3}$ A; (b) 75 A; (c) $\frac{1}{3}$ A; (d) 3 A.

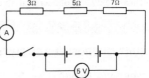

Fig 21

4 The effect of connecting an additional parallel load to an electrical supply source is to increase the
 (a) resistance of load;
 (b) voltage of the source;
 (c) current taken from the source;
 (d) p.d. across the load.

5 The equivalent resistance when a resistor of $\frac{1}{3}$ Ω is connected in parallel with a $\frac{1}{4}$ Ω resistor is (a) $\frac{1}{7}$ Ω; (b) 7 Ω.

6 Which of the following formulae for electrical power is incorrect?
 (a) VI; (b) $\frac{V}{I}$; (c) I^2R; (d) $\frac{V^2}{R}$.

7 The power dissipated by a resistor of 4 Ω when a current of 5 A passes through it is
 (a) 6.25 W, (b) 20 W, (c) 80 W, (d) 100 W.

8 Which of the following statements is true?
 (a) Series-connection of lamps is usually employed in house-wiring.
 (b) 200 kΩ resistance is equivalent to 0.02 MΩ.
 (c) An ammeter has a low resistance and must be connected in parallel with a circuit.
 (d) An electrical insulator has a high resistance.

9 A current of 3 A flows for 50 h through a 6 Ω resistor. The energy consumed by the resistor is:
 (a) 0.9 kWh; (b) 2.7 kWh; (c) 9 kWh; (d) 27 kWh.

10 What must be known in order to calculate the energy used by an electrical appliance?
 (a) voltage and current;
 (b) current and time of operation;
 (c) power and time of operation;
 (d) current and resistance.

(c) CONVENTIONAL PROBLEMS

1 The current flowing through a heating element is 5 A when a p.d. of 35 V is applied across it. Find the resistance of the element. [7 Ω]

2 Determine the p.d. which must be applied to a 5 kΩ resistor such that a current of 6 mA may flow. [30 V]

3 The hot resistance of a 250 V filament lamp is 625 Ω. Determine the current taken by the lamp and its power rating. [0.4 A; 100 W]

4 Determine the resistance of a coil connected to a 150 V supply when a current of (a) 75 mA; (b) 300 μA flows through it. [(a) 2 kΩ; (b) 0.5 MΩ]

5 Determine the resistance of an electric fire which takes a current of 12 A from a 240 V supply. Find also the power rating of the fire and the energy used in 20 h.
[20 Ω; 2.88 kW; 57.6 kWh]

6 The p.d.'s measured across three resistors connected in series are 5 V, 7 V and 10 V, and the supply current is 2 A. Determine (a) the supply voltage, (b) the total circuit resistance and (c) the values of the three resistors.
[(a) 22 V; (b) 11 Ω; (c) 2.5 Ω, 3.5 Ω, 5 Ω]

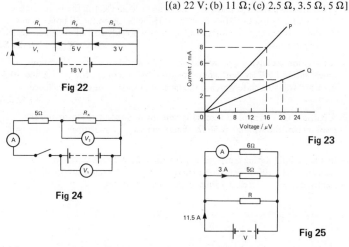

7 For the circuit shown in *Fig 22* determine the value of V_1. If the total circuit resistance is 36 Ω determine the supply current and the value of the resistors R_1, R_2 and R_3. [10 V; 0.5 A; 20 Ω, 10 Ω, 6 Ω]

8 Graphs of current against voltage for two resistors P and Q are shown in *Fig 23*. Determine the value of each resistor. [2 mΩ; 5 mΩ]

9 When the switch in the circuit in *Fig 24* is closed the reading on voltmeter 1 is 30 V and that on voltmeter 2 is 10 V. Determine the reading on the ammeter and the value of resistor R_x. [4 A; 2.5 Ω]

10 Two resistors are connected in series across an 18 V supply and a current of 5 A flows. If one of the resistors has a value of 2.4 Ω determine (a) the value of the other resistor and (b) the p.d. across the 2.4 Ω resistor. [(a) 1.2 Ω; (b) 12 V]

11 Resistances of 4 Ω and 12 Ω are connected in parallel across a 9 V battery. Determine (a) the equivalent circuit resistance, (b) the supply current, and (c) the current in each resistor. [(a) 3 Ω; (b) 3 A; (c) $2\frac{1}{4}$ A; $\frac{3}{4}$ A]

12 For the circuit shown in *Fig 25* determine (a) the reading on the ammeter, and
(b) the value of resistor R. [2.5 A; 2.5 Ω]

13 If four identical lamps are connected in parallel and the combined resistance is
100 Ω, find the resistance of one lamp. [400 Ω]

14 Three identical filament lamps are connected (a) in series, (b) in parallel across a
210 V supply. State for each connection the p.d. across each lamp.
[(a) 70 V; (b) 210 V]

15 A 60 W electric light bulb is connected to a 240 V supply. Determine (a) the
current flowing in the bulb and (b) the resistance of the bulb.
[(a) 0.25 A; (b) 960 Ω]

16 Determine the power dissipated when a current of 10 mA flows through an
appliance having a resistance of 8 kΩ. [0.8 W]

17 Calculate the power dissipated by the element of an electric fire of resistance
30 Ω when a current of 10 A flows in it. If the fire is on for 30 hours in a week
determine the energy used. Determine also the weekly cost of energy if electricity
costs 5p per unit. [3 kW, 90 kWh, £4.50]

18 A television set having a power rating of 120 W and an electric lawnmower of
power rating 1 kW are both connected to a 240 V supply. If 3 A, 5 A and 10 A
fuses are available state which is the most appropriate for each appliance.
[3 A; 5 A]

19 A p.d. of 500 V is applied across the winding of an electric motor and the
resistance of the winding is 50 Ω. Determine the power dissipated by the coil.
[5 kW]

20 In a household during a particular week three 2 kW fires are used on average 25 h
each and eight 100 W lights are used on average 35 h each. Determine the cost of
electricity for the week if 1 unit of electricity costs 5p. [£8.90]

15 Resistance variation and electromagnetism

A. MAIN POINTS CONCERNED WITH RESISTANCE VARIATION AND ELECTROMAGNETISM

Resistance variation

1. The resistance of an electrical conductor depends on 4 factors, these being:
   (a) the length of the conductor, (b) the cross-sectional area of the conductor, (c) the type of material and (d) the temperature of the material.

2. (i) Resistance, R, is directly proportional to length, l, of a conductor, i.e. $R \propto l$. Thus, for example, if the length of a piece of wire is doubled, then the resistance is doubled.

   (ii) Resistance, R, is inversely proportional to cross-sectional area, a, of a conductor, i.e. $R \propto \frac{1}{a}$. Thus, for example, if the cross-sectional area of a piece of wire is doubled then the resistance is halved.

   (iii) Since $R \propto l$ and $R \propto \frac{1}{a}$ then $R \propto \frac{l}{a}$. By inserting a constant of proportionality into this relationship the type of material used may be taken into account. The constant of proportionality is known as the **resistivity** of the material and is given the symbol ρ (rho).
   Thus,
   $$\text{resistance} \quad R = \frac{\rho l}{a} \text{ ohms}$$
   ρ is measured in ohm metres (Ω m).
   The value of the resistivity is that resistance of a unit cube of the material measured between opposite faces of the cube.

   (iv) Resistivity varies with temperature and some typical values of resistivities measured at about room temperature are given below:

   Copper 1.7×10^{-8} Ωm (or 0.017 $\mu\Omega$m)
   Aluminium 2.6×10^{-8} Ωm (or 0.026 $\mu\Omega$m)
   Carbon (graphite) 10×10^{-8} Ωm (0.10 $\mu\Omega$m)
   Glass 1×10^{10} Ωm (or 10^4 $\mu\Omega$m)
   Mica 1×10^{13} Ωm (or 10^7 $\mu\Omega$m)

   Note that good conductors of electricity have a low value of resistivity and good insulators have a high value of resistivity.
   (See *Problems 1 to 7*)

3. (i) In general, as the temperature of a material increases, most conductors increase in resistance, insulators decrease in resistance whilst the resistance of some special alloys remain almost constant.

   (ii) The **temperature coefficient of resistance** of a material is the increase in the resistance of a 1 Ω resistor of that material when it is subjected to a rise of temperature of 1°C. The symbol used for the temperature coefficient of resistance is α (alpha). Thus, if some copper wire of resistance 1 Ω is heated through 1°C and its resistance is then measured as 1.0043 Ω then $\alpha = 0.0043$ Ω/Ω °C for copper. The units are usually expressed only as 'per °C', i.e., $\alpha = 0.0043/$°C for copper. If the 1 Ω resistor of copper is heated through 100°C then the resistance at 100°C would be $1 + 100 \times 0.0043 = 1.43$ Ω.

   (iii) If the resistance of a material at 0°C is known the resistance at any other temperature can be determined from:

   $$R_\theta = R_0(1 + \alpha_0 \theta)$$

   where R_0 = resistance at 0°C;
   R_θ = resistance at temperature θ°C;
   α_0 = temperature coefficient of resistance at 0°C.

   (iv) If the resistance at 0°C is not known, but is known at some other temperature θ_1, then the resistance at any temperature can be found as follows:

   $$R_1 = R_0(1 + \alpha_0 \theta_1) \text{ and } R_2 = R_0(1 + \alpha_0 \theta_2)$$

   Dividing one equation by the other gives:

   $$\frac{R_1}{R_2} = \frac{1 + \alpha_0 \theta_1}{1 + \alpha_0 \theta_2}$$

   where R_2 = resistance at temperature θ_2.

   (v) If the resistance of a material at room temperature (approximately 20°C), R_{20}, and the temperature coefficient of resistance at 20°C, α_{20} are known then the resistance R_θ at temperature θ°C is given by:

   $$R_\theta = R_{20}[1 + \alpha_{20}(\theta - 20)]$$

   (vi) Some typical values of temperature coefficient of resistance measured at 0°C are given below:

Copper	0.0043/°C	Aluminium	0.0038/°C
Nickel	0.0062/°C	Carbon	−0.00048/°C
Constantan	0	Eureka	0.00001/°C

   (Note that the negative sign for carbon indicates that its resistance falls with increase of temperature.)

   (See *Problems 8 to 13*)

Electromagnetism

4. A **permanent magnet** is a piece of ferromagnetic material (such as iron, nickel or cobalt) which has properties of attracting other pieces of these materials.

5. The area around a magnet is called the **magnetic field** and it is in this area that the effects of the **magnetic force** produced by the magnet can be detected.

6. The magnetic field of a bar magnet can be represented pictorially by the 'lines of force' (or lines of 'magnetic flux' as they are called) as shown in *Fig 1*. Such a field pattern can be produced by placing iron filings in the vicinity of the magnet. The

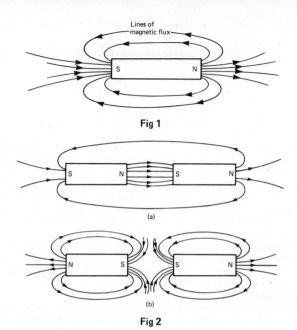

Fig 1

Fig 2

field direction at any point is taken as that in which the north-seeking pole of a compass needle points when suspended in the field. External to the magnet the direction of the field is north to south.

7 The laws of magnetic attraction and repulsion can be demonstrated by using two bar magnets. In *Fig 2(a)*, with **unlike poles** adjacent, **attraction** occurs. In *Fig 2(b)*, with **like poles** adjacent, **repulsion** occurs.

8 Magnetic fields are produced by electric currents as well as by permanent magnets. The field forms a circular pattern with the current carrying conductor at the centre. The effect is portrayed in *Fig 3* where the convention adopted is:
   (a) current flowing **away** from the viewer is shown by ⊕ —can be thought of as the feathered end of the shaft of an arrow
   (b) current flowing **towards** the viewer is shown by ⊙ —can be thought of as the tip of an arrow.

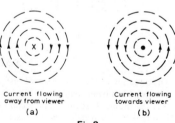

Fig 3

9 The **direction** of the fields in *Fig 3* is remembered by the **screw rule** which states: 'If a normal right-hand thread screw is screwed along the conductor in the direction of the current, the direction of rotation of the screw is in the direction of the magnetic field'.

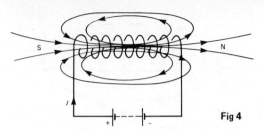

Fig 4

10 A magnetic field produced by a long coil, or **solenoid**, is shown in *Fig 4* and is seen to be similar to that of a bar magnet shown in *Fig 1*. If the solenoid is wound on an iron bar an even stronger field is produced. The **direction** of the field produced by current *I* is determined by a compass and is remembered by either:
  (a) the **screw rule**, which states that if a normal right hand thread screw is placed along the axis of the solenoid and is screwed in the direction of the current it moves in the direction of the magnetic field inside of the solenoid (i.e., points in the direction of the north pole), or
  (b) the **grip rule**, which states that if the coil is gripped with the right hand with the fingers pointing in the direction of the current, then the thumb, outstretched parallel to the axis of the solenoid, points in the direction of the magnetic field inside the solenoid (i.e. points in the direction of the north pole).

(See *Problem 14*)

11 An **electromagnet**, which is a solenoid wound on an iron core, provides the basis of many items of electrical equipment, examples including electric bells, relays, lifting magnets and telephone receivers. (See *Problems 15 to 17*.)

12 If the current-carrying conductor shown in *Fig 3(a)* is placed in the magnetic field shown in *Fig 5(a)* then the two fields interact and cause a force to be exerted on the conductor as shown in *Fig 5(b)*. The field is strengthened above the conductor and weakened below, thus tending to move the conductor downwards. This is the basic principle of operation of the **electric motor** (see *Problem 18*) and the **moving-coil instrument** (see *Problem 19*).

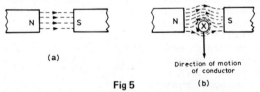

Fig 5

13 The direction of the force exerted on a conductor can be predetermined by using **Fleming's left-hand rule** (often called the motor rule), which states:
'Let the thumb, first finger and second finger of the left hand be extended such that they are all at right angles to each other, as shown in *Fig 6*. If the first finger points in the direction of the magnetic field, the second finger points in the direction of the current, then the thumb will point in the direction of the motion of the conductor.'

Summarising: First finger   – <u>F</u>ield
                Se<u>C</u>ond finger – <u>C</u>urrent
                Thu<u>M</u>b   – <u>M</u>otion

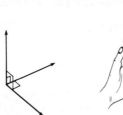
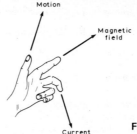

Fig 6

14 **Electromagnetic induction** is the production of voltage in a conductor which is in the region of a changing magnetic field (see *Problem 20*). Applications of electromagnetic induction include **a.c. generators** (see *Problem 21*) and transformers.

Electricity is produced by generators at power stations and then distributed by a vast network of transmission lines (called the National Grid system) to industry and for domestic use. It is easier and cheaper to generate alternating current (a.c.) than direct current (d.c.) and a.c. is more conveniently distributed than d.c. since its voltage can be readily altered using transformers.

B. WORKED PROBLEMS ON RESISTANCE VARIATION AND ELECTRO-MAGNETISM

Problem 1 The resistance of a 5 m length of wire is 600 Ω. Determine (a) the resistance of an 8 m length of the same wire, and (b) the length of the same wire when the resistance is 420 Ω.

(a) Resistance, R, is directly proportional to length, l, i.e. $R \propto l$.
Hence, 600 $\Omega \propto$ 5 m or 600 = $(k)(5)$,
where k is the coefficient of proportionality.

Hence, $k = \dfrac{600}{5} = 120$

When the length l is 8 m, then
resistance $R = kl = (120)(8) = \mathbf{960\ \Omega}$

(b) When the resistance is 420 Ω, 420 = kl, from which,
length $l = \dfrac{420}{k} = \dfrac{420}{120} = \mathbf{3.5\ m}$

Problem 2 A piece of wire of cross-sectional area 2 mm^2 has a resistance of 300 Ω. Find (a) the resistance of a wire of the same length and material if the cross-sectional area is 5 mm^2, (b) the cross-sectional area of a wire of the same length and material of resistance 750 Ω.

Resistance R is inversely proportional to cross-sectional area, a, i.e. $R \propto \dfrac{1}{a}$
Hence 300 $\Omega \propto \dfrac{1}{2\ \text{mm}^2}$ or 300 = $(k)(\dfrac{1}{2})$, from which, the coefficient of proportionality, $k = 300 \times 2 = 600$

(a) When the cross-sectional area $a = 5$ mm² then $R = (k)(\frac{1}{5}) = (600)(\frac{1}{5}) = 120\ \Omega$

(Note that resistance has decreased as the cross-sectional is increased.)

(b) When the resistance is 750 Ω then $750 = (k)(\frac{1}{a})$, from which,

cross-sectional area, $a = \dfrac{k}{750} = \dfrac{600}{750} = 0.8$ mm²

Problem 3 A wire of length 8 m and cross-sectional area 3 mm² has a resistance of 0.16 Ω. If the wire is drawn out until its cross-sectional area is 1 mm², determine the resistance of the wire.

Resistance R is directly proportional to length, l, and inversely proportional to the cross-sectional area, i.e.,

i.e., $R \propto \dfrac{l}{a}$ or $R = k\dfrac{l}{a}$, where k is the coefficient of proportionality.

Since $R = 0.16$, $l = 8$ and $a = 3$, then $0.16 = (k)(\frac{8}{3})$, from which $k = 0.16 \times \frac{3}{8} = 0.06$.

If the cross-sectional area is reduced to $\frac{1}{3}$ of its original area then the length must be tripled to 3×8, i.e., 24 m

New resistance $R = k\dfrac{l}{a} = 0.06\ (\frac{24}{1}) = 1.44\ \Omega$

Problem 4 Calculate the resistance of a 2 km length of aluminium overhead power cable if the cross-sectional area of the cable is 100 mm². Take the resistivity of aluminium to be 0.03×10^{-6} Ωm.

Length $l = 2$ km $= 2000$ m; area, $a = 100$ mm² $= 100 \times 10^{-6}$ m²; resistivity $\rho = 0.03 \times 10^{-6}$ Ωm

Resistance $R = \dfrac{\rho l}{a} = \dfrac{(0.03 \times 10^{-6}\ \Omega m)(2000\ m)}{(100 \times 10^{-6}\ m^2)} = \dfrac{0.03 \times 2000}{100}\ \Omega = 0.6\ \Omega$

Problem 5 Calculate the cross-sectional area, in mm², of a piece of copper wire, 40 m in length and having a resistance of 0.25 Ω. Take the resistivity of copper as 0.02×10^{-6} Ωm.

Resistance $R = \dfrac{\rho l}{a}$ hence cross-sectional area $a = \dfrac{\rho l}{R} = \dfrac{(0.02 \times 10^{-6}\ \Omega m)(40\ m)}{0.25\ \Omega}$

$= 3.2 \times 10^{-6}$ m²
$= (3.2 \times 10^{-6}) \times 10^6$ mm² $= 3.2$ mm²

Problem 6 The resistance of 1.5 km of wire of cross-sectional area 0.17 mm² is 150 Ω. Determine the resistivity of the wire.

Resistance $R = \dfrac{\rho l}{a}$

Hence,
the resistivity $\rho = \dfrac{Ra}{l} = \dfrac{(150\ \Omega)(0.17 \times 10^{-6}\ m^2)}{(1500\ m)}$

$= 0.017 \times 10^{-6}$ Ωm or $0.017\ \mu\Omega m$

Problem 7 Determine the resistance of 1200 m of copper cable having a diameter of 12 mm if the resistivity of copper is 1.7×10^{-8} Ωm.

Cross-sectional area of cable, $a = \pi r^2 = \pi \left(\dfrac{12}{2}\right)^2 = 36\pi$ mm$^2 = 36\pi \times 10^{-6}$ m^2

Resistance $R = \dfrac{\rho l}{a} = \dfrac{(1.7 \times 10^{-8}\ \Omega\text{m})(1200\ \text{m})}{(36\pi \times 10^{-6}\ \text{m}^2)} = \dfrac{1.7 \times 1200 \times 10^6\ \Omega}{10^8 \times 36\pi}$

$= \dfrac{1.7 \times 12}{36\pi}\ \Omega = \mathbf{0.180\ \Omega}$

Problem 8 A coil of copper wire has a resistance of 100 Ω when its temperature is 0°C. Determine its resistance at 100°C if the temperature coefficient of resistance of copper at 0°C is 0.0043/°C.

Resistance $R_\theta = R_0(1 + \alpha_0 \theta)$ from para. 3(iii).
Hence resistance at 100°C, $R_{100} = 100\ [1 + (0.0043)(100)]$
$= 100\ [1 + 0.43] = 100\ (1.43) = \mathbf{143\Omega}$

Problem 9 An aluminium cable has a resistance of 27 Ω at a temperature of 35°C. Determine its resistance at 0°C. Take the temperature coefficient of resistance at 0°C to be 0.0038/°C.

Resistance at θ°C, $R_\theta = R_0\ (1 + \alpha_0 \theta)$

Hence resistance at 0°C, $R_0 = \dfrac{R_\theta}{(1 + \alpha_0 \theta)} = \dfrac{27}{[1 + (0.0038)(35)]}$

$= \dfrac{27}{1 + 0.133} = \dfrac{27}{1.133} = \mathbf{23.83\ \Omega}$

Problem 10 A carbon resistor has a resistance 1 kΩ at 0°C. Determine its resistance at 80°C. Assume that the temperature coefficient of resistance for carbon at 0°C is -0.0005.

Resistance at temperature θ°C, $R_\theta = R_0\ (1 + \alpha_0 \theta)$
i.e. $R_\theta = 1000\ [1 + (-0.0005)(80)]$
$= 1000\ [1 - 0.040] = 1000(0.96) = \mathbf{960\ \Omega}$

Problem 11 A coil of copper wire has a resistance of 10 Ω at 20°C. If the temperature coefficient of resistance of copper at 20°C is 0.004/°C determine the resistance of the coil when the temperature rises to 100°C.

Resistance at θ°C, $R_\theta = R_{20}[1 + \alpha_{20}(\theta - 20)]$, from para. 3(v)

Hence resistance at 100°C, $R_{100} = 10\ [1 + (0.004)(100 - 20)]$

$= 10\ [1 + (0.004)(80)]$

$= 10\ [1 + 0.32]$

$= 10(1.32) = \mathbf{13.2\ \Omega}$

Problem 12 The resistance of a coil of aluminium wire at 18°C is 200 Ω. The temperature of the wire is increased and the resistance rises to 240 Ω. If the temperature coefficient of resistance of aluminium is 0.0039/°C at 18°C determine the temperature to which the coil has risen.

Let the temperature rise to θ°C.
Resistance at θ°C, $R_\theta = R_{18}[1 + \alpha_{18}(\theta - 18)]$

i.e. $240 = 200[1 + (0.0039)(\theta - 18)]$

$240 = 200 + (200)(0.0039)(\theta - 18)$

$240 - 200 = 0.78(\theta - 18)$

$40 = 0.78(\theta - 18)$

$\dfrac{40}{0.78} = \theta - 18$

$51.28 = \theta - 18$, from which, $\theta = 51.28 + 18 = 69.28$°C

Hence the temperature of the coil increases to 69.28°C

Problem 13 Some copper wire has a resistance of 200 Ω at 20°C. A current is passed through the wire and the temperature rises to 90°C. Determine the resistance of the wire at 90°C, correct to the nearest ohm, assuming that the temperature coefficient of resistance is 0.004/°C at 0°C.

$R_{20} = 200$ Ω, $\alpha_0 = 0.004$/°C

$\dfrac{R_{20}}{R_{90}} = \dfrac{[1 + \alpha_0(20)]}{[1 + \alpha_0(90)]}$, from para. 3(iv).

Hence $R_{90} = \dfrac{R_{20}[1 + 90\alpha_0]}{[1 + 20\alpha_0]} = \dfrac{200[1 + 90(0.004)]}{[1 + 20(0.004)]} = \dfrac{200[1 + 0.36]}{[1 + 0.08]}$

i.e. $= \dfrac{200(1.36)}{(1.08)} = 251.85$ Ω

i.e. the resistance of the wire at 90°C is 252 Ω.

Problem 14 Fig 7 shows a coil of wire wound on an iron core connected to a battery. Sketch the magnetic field pattern associated with the current carrying coil and determine the polarity of the field.

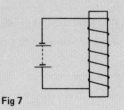

Fig 7

The magnetic field associated with the solenoid of *Fig 7* is similar to the field associated with a bar magnet and is as shown in *Fig 8*. The polarity of the field is determined either by the screw rule or by the grip rule as explained in para. 10. Thus the north pole is at the bottom and the south pole at the top.

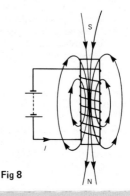

Fig 8

Problem 15 State what happens when the switch S is closed and then opened in the circuit shown in *Fig 9*.

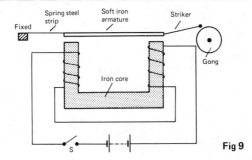

Fig 9

When the switch S is closed a current passes through the coil. The iron-cored solenoid is energised, the soft iron armature is attracted to the electromagnet and the striker hits the gong. When the switch S is opened the coil becomes demagnetised and the spring steel strip pulls the armature back to its original position. This is the principle of operation of an electric bell.

Problem 16 For the relay circuit shown in *Fig 10* state what happens when the switch S is closed.

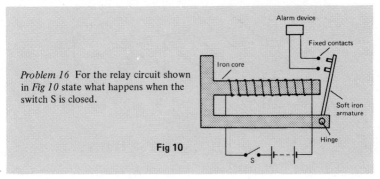

Fig 10

When the switch S is closed a current passes through the coil and the iron-cored solenoid is energised. The hinged soft iron armature is attracted to the electromagnet and pushes against the two fixed contacts so that they are connected together, thus closing the electric circuit to be controlled—in this case, an alarm circuit. The alarm sounds for as long as the current flows in the coil.

Problem 17 Make a sketch showing the main parts of a lifting magnet that would be suitable for use in a scrap-metal yard and state its principle of operation.

A typical lifting magnet showing the plan and elevation is shown in *Fig 11*. When current is passed through the coil, the iron core becomes magnetised,(i.e. an electromagnet) and thus will attract to it other pieces of magnetic material. When

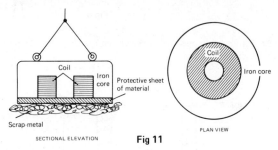

Fig 11

the circuit is broken the iron core becomes demagnetised which releases the materials being lifted.

Problem 18 Briefly describe, with an appropriate sketch, the principle of operation of a simple d.c. motor.

A rectangular coil which is free to rotate about a fixed axis is shown placed inside a magnetic field produced by permanent magnets in *Fig 12*. A direct current is fed into the coil via carbon brushes bearing on a commutator, which consists of a

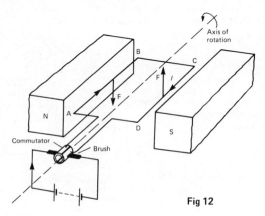

Fig 12

metal ring split into two halves separated by insulation. When current flows in the coil a magnetic field is set up around the coil which interacts with the magnetic field produced by the magnets. This causes a force F to be exerted on the current carrying conductor, which, by Fleming's left-hand rule (see para. 13) is downwards between points A and B and upwards between C and D for the current direction shown. This causes a torque and the coil rotates anticlockwise.

When the coil has turned through 90° from the position shown in *Fig 12* the brushes connected to the positive and negative terminals of the supply make contact with different halves of the commutator ring, thus reversing the direction of the current flow in the conductor. If the current is not reversed and the coil rotates past this position the forces acting on it change direction and it rotates in the opposite direction thus never making more than half a revolution. The current direction is reversed every time the coil swings through the vertical position and thus the coil rotates anticlockwise for as long as the current flows. This is the principle of operation of a d.c. motor which is thus a device that takes in electrical energy and converts it into mechanical energy.

Problem 19 Describe, with the aid of diagrams, the principle of operation of a moving-coil instrument.

A moving coil instrument operates on the motor principle, i.e. when a conductor carrying current is placed in a magnetic field a force is exerted on the conductor. In a moving-coil instrument a coil is placed centrally in the gap between shaded pole pieces as shown by the front elevation in *Fig 13(a)*. The airgap is kept as small as possible, but for clarity is shown larger in size.

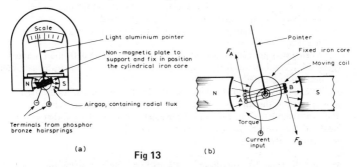

Fig 13

The coil is supported by steel pivots, resting in jewel bearings, on a cylindrical iron core. Current is led into and out of the coil by two phosphor bronze spiral hairsprings which are wound in opposite directions to minimise the effect of temperature change and to limit the coil swing (i.e. to control the movement) and return the movement to zero position when no current flows.

Current flowing in the coil produces forces as shown in *Fig 13(b)*, the directions being obtained by Fleming's left-hand rule (see para. 13). The two forces, F_A and F_B, produce a torque which will move the coil in a clockwise direction, i.e., move the pointer from left to right. The force is proportional to current and the scale is linear. The moving coil instrument will only measure direct current or voltage and the terminals are marked positive and negative to ensure that the current passes through the coil in the correct direction to deflect the pointer 'up the scale'.

Problem 20 Briefly describe electromagnetic induction with reference to the movement of a magnet in a coil connected to a meter.

Fig 14(a) shows a coil of wire connected to a centre-zero galvanometer, which is a sensitive ammeter with the zero-current position in the centre of the scale.
(a) When the magnet is moved at constant speed towards the coil (*Fig 14(a)*), a deflection is noted on the galvanometer showing that a current has been produced in the coil.
(b) When the magnet is moved at the same speed as in (a) but away from the coil the same deflection is noted but is in the opposite direction (see *Fig 14(b)*).
(c) When the magnet is held stationary even within the coil no deflection is recorded.

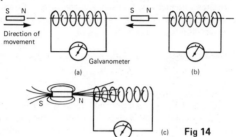

Fig 14

(d) When the coil is moved at the same speed as in (a) and the magnet held stationary the same galvanometer deflection is noted.
(e) When the relative speed is, say, doubled, the galvanometer deflection is doubled.
(f) When a stronger magnet is used, a greater galvanometer deflection is noted.
(g) When the number of turns of wire of the coil is increased, a greater galvanometer deflection is noted.

Fig 14(c) shows the magnetic field associated with the magnet. As the magnet is moved towards the coil, the magnetic flux of the magnet moves across, or cuts, the coil. **It is the relative movement of the magnetic flux and the coil that causes an emf, and thus current, to be induced in the coil.** This effect is known as electromagnetic induction.

Problem 21 Describe, with diagrams, the basic principle of operation of an a.c. generator.

Let a single turn coil be free to rotate at constant speed symmetrically between the poles of a magnet system as shown in *Fig 15*. By electromagnetic induction (see *Problem 20*), an emf is generated in the coil. The magnitude of the emf varies and reverses its direction at regular intervals. The reason for this is shown in *Fig 16*.

In positions (a), (e) and (i) the conductors of the loop are effectively moving

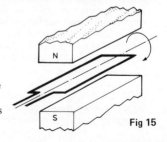

Fig 15

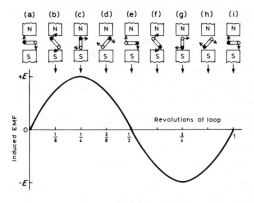

Fig 16

along the magnetic field, no flux is cut and hence no emf is induced. In position (c), maximum flux is cut and hence maximum emf is induced. In position (g), maximum flux is cut and hence maximum emf is again induced. However, the flux is cut in the opposite direction to that in position (c) and is thus shown as maximum negative in *Fig 16*. In positions (b), (d), (f) and (h) some flux is cut and hence some emf is induced. If all such positions of the coil are considered, in one revolution of the coil, one cycle of alternating emf is produced as shown. This is the principle of operation of the a.c. generator or the alternator.

C. FURTHER PROBLEMS ON RESISTANCE VARIATION AND ELECTROMAGNETISM

(a) SHORT ANSWER PROBLEMS

1   Name four factors which can affect the resistance of a conductor.

2   If the length of a piece of wire of constant cross-sectional area is halved, the resistance of the wire is . . . .

3   If the cross-sectional area of a certain length of cable is trebled, the resistance of the cable is . . . .

4   What is resistivity? State its unit and the symbol used.

5   Complete the following:
Good conductors of electricity have a . . . . value of resistivity and good insulators have a . . . . value of resistivity.

6   What is meant by the 'temperature coefficient of resistance'?
State its units and the symbols used.

7   If the resistance of a metal at 0°C is R_0, R_θ is the resistance at θ°C and α_0 is the temperature coefficient of resistance at 0°C then: $R_\theta = $ . . . .

8   What is a permanent magnet?

9   Sketch the pattern of the magnetic field associated with a bar magnet. Mark the direction of the field.

10 The direction of the magnetic field around a current carrying conductor may be remembered using the .... rule.

11 Sketch the magnetic field pattern associated with a solenoid connected to a battery and wound on an iron bar. Show the direction of the field.

12 Name three applications of electromagnetism.

13 State what happens when a current carrying conductor is placed in a magnetic field between two magnets.

14 The direction of the force exerted on a current carrying conductor in a magnetic field can be predetermined using .... rule.

15 What is electromagnetic induction?

16 Name two applications of electromagnetic induction.

17 Briefly explain the principle of operation of a moving coil ammeter.

18 Explain the principle of operation of a d.c. motor.

19 What happens when a permanent magnet is moved in a coil of wire connected to a galvanometer.

20 Briefly describe the principle of operation of an a.c. generator.

(b) MULTI-CHOICE PROBLEMS (answers on page 149)

1 The length of a certain conductor of resistance 100 Ω is doubled and its cross-sectional area is halved. Its new resistance is:
(a) 100 Ω; (b) 200 Ω; (c) 50 Ω; (d) 400 Ω.

2 The resistance of a 2 km length of cable of cross-sectional area 2 mm^2 and resistivity of 2×10^{-8} Ωm is:
(a) 0.02 Ω; (b) 20 Ω; (c) 0.02 mΩ; (d) 200 Ω.

3 Which of the following statements is true?
(a) A good conductor of electricity has a high resistivity.
(b) When the temperature of carbon is increased its resistance increases.
(c) A moving-coil meter operates on the motor principle.
(d) Fleming's left-hand rule is often called the generator rule.

4 A coil of wire has a resistance of 10 Ω at 0°C. If the temperature coefficient of resistance for the wire is 0.004/°C its resistance at 100°C is:
(a) 0.4 Ω; (b) 1.4 Ω; (c) 14 Ω; (d) 10 Ω.

5 Three bar magnets are arranged as shown in *Fig 17* and their lines of flux indicated. The polarities of the magnets can be as follows:

	Magnet P	Magnet Q	Magnet R
(a)	N	N	S
(b)	N	S	S
(c)	N	S	N
(d)	S	N	N

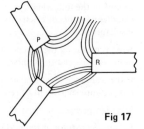

Fig 17

6 If a conductor is horizontal, the current flowing from left to right and the

direction of the surrounding magnetic field is from above to below, the force exerted on the conductor is:
(a) from left to right;
(b) from below to above;
(c) away from the viewer;
(d) towards the viewer.

7 For the current carrying conductor lying in the magnetic field shown in *Fig 18*, the direction of the force on the conductor is:
(a) to the left; (b) upwards; (c) to the right; (d) downwards.

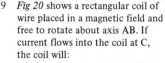

Fig 18

8 For the current carrying conductor lying in the magnetic field shown in *Fig 19*, the direction of the current in the conductor is:
(a) towards the viewer; (b) away from the viewer.

9 *Fig 20* shows a rectangular coil of wire placed in a magnetic field and free to rotate about axis **AB**. If current flows into the coil at C, the coil will:
(a) commence to rotate anticlockwise;
(b) commence to rotate clockwise;
(c) remain in the vertical position;
(d) experience a force towards the north pole.

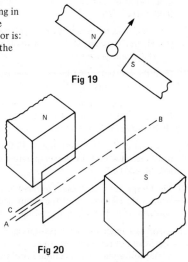

Fig 19

Fig 20

10 A bar magnet is moved at a steady speed of 1.0 m/s towards a coil of wire which is connected to a centre-zero galvanometer. The magnet is now withdrawn along the same path at 0.5 m/s. The deflection of the galvanometer:
(a) is in the same direction as previously with the magnitude of the deflection doubled;
(b) is in the opposite direction as previous with the magnitude of the deflection halved;
(c) is in the same direction as previous with the magnitude of the deflection halved;
(d) is in the opposite direction as previous with the magnitude of the deflection doubled.

(c) CONVENTIONAL PROBLEMS

1 The resistance of a 2 m length of cable is 2.5 Ω. Determine (a) the resistance of a 7 m length of the same cable and (b) the length of the same wire when the resistance is 6.25 Ω. [(a) 8.75 Ω; (b) 5 m]

2 Some wire of cross-sectional area 1 mm² has a resistance of 20 Ω. Determine (a) the resistance of a wire of the same length and material if the cross-sectional

area is 4 mm², and (b) the cross-sectional area of a wire of the same length and material if the resistance is 32 Ω. [(a) 5 Ω; (b) 0.625 mm²]

3 Some wire of length 5 m and cross-sectional area 2 mm² has a resistance of 0.08 Ω. If the wire is drawn out until its cross-sectional area is 1 mm², determine the resistance of the wire. [0.32 Ω]

4 Find the resistance of 800 m of copper cable of cross-sectional area 20 mm². Take the resistivity of copper as 0.02 μΩm. [0.8 Ω]

5 Calculate the cross-sectional area, in mm², of a piece of aluminium wire 100 m long and having a resistance of 2 Ω. Take the resistivity of aluminium as 0.03×10^{-6} Ωm. [1.5 mm²]

6 (a) What does the resistivity of a material mean?
   (b) The resistance of 500 m of wire of cross-sectional area 2.6 mm² is 5 Ω. Determine the resistivity of the wire in μΩm. [0.026 μΩm]

7 Find the resistance of 1 km of copper cable having a diameter of 10 mm if the resistivity of copper is 0.017×10^{-6} Ωm. [0.216 Ω]

8 A coil of aluminium wire has a resistance of 50 Ω when its temperature is 0°C. Determine its resistance at 100°C if the temperature coefficient of resistance of aluminium at 0°C is 0.0038/°C. [69 Ω]

9 A copper cable has a resistance of 30 Ω at a temperature of 50°C. Determine its resistance at 0°C. Take the temperature coefficient of resistance of copper at 0°C as 0.0043/°C. [24.69 Ω]

10 The temperature coefficient of resistance for carbon at 0°C is −0.00048/°C. What is the significance of the minus sign? A carbon resistor has a resistance of 500 Ω at 0°C. Determine its resistance at 50°C. [488 Ω]

11 A coil of copper wire has a resistance of 20 Ω at 18°C. If the temperature coefficient of resistance of copper at 18°C is 0.004/°C, determine the resistance of the coil when the temperature rises to 98°C. [26.4 Ω]

12 The resistance of a coil of nickel wire at 20°C is 100 Ω. The temperature of the wire is increased and the resistance rises to 130 Ω. If the temperature coefficient of resistance of nickel is 0.006/°C at 20°C, determine the temperature to which the coil has risen. [70°C]

13 Some aluminium wire has a resistance of 50 Ω at 20°C. The wire is heated to a temperature of 100°C. Determine the resistance of the wire at 100°C, assuming that the temperature coefficient of resistance at 0°C is 0.004/°C. [64.8 Ω]

14 A copper cable is 1.2 km long and has a cross-sectional area of 5 mm². Find its resistance at 80°C if at 20°C the resistivity of copper is 0.02×10^{-6} Ωm and its temperature coefficient of resistance is 0.004/°C. [5.952 Ω]

15 *Fig 21* shows a simplified diagram of a section through the coil of a moving-coil instrument. For the direction of current flow shown in the coils determine the direction that the pointer will move.

[to the left]

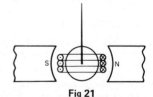

Fig 21

16 With the aid of sketches, describe the type of magnetic field produced by (a) a bar magnet and (b) a solenoid.

17 Describe what happens when a current-carrying conductor is placed in a magnetic field. Hence explain, with suitable diagrams, the basic operation of (a) a moving coil meter, and (b) a d.c. motor.

18 A coil of wire is connected to a centre-zero galvanometer. State what deflection is noted on the galvanometer when a permanent magnet is (i) moved at a steady speed towards the coil, (ii) moved at the same speed away from the coil, (iii) held stationary in the coil and (iv) moved at half the original speed towards the coil.

19 Describe the principle of electromagnetic induction and explain how it may be used to produce an alternating current.

16 Chemical effects of electricity

A. MAIN POINTS CONCERNED WITH THE CHEMICAL EFFECTS OF ELECTRICITY

1 A material must contain **charged particles** to be able to conduct electric current. In **solids**, the current is carried by **electrons**. Copper, lead, aluminium, iron and carbon are some examples of solid conductors. In **liquids and gases**, the current is carried by the part of a molecule which has acquired an electric charge, called **ions**. These can possess a positive or negative charge, and examples include hydrogen ion H^+, copper ion Cu^{++} and hydroxyl ion OH^-. Distilled water contains no ions and is a poor conductor of electricity whereas salt water contains ions and is a fairly good conductor of electricity.

2 (i) **Electrolysis** is the decomposition of a liquid compound by the passage of electric current through it. Practical applications of electrolysis include the electroplating of metals (see para. 3), the refining of copper and the extraction of aluminium from its ore.

   (ii) An **electrolyte** is a compound which will undergo electrolysis. Examples include salt water, copper sulphate and sulphuric acid.

   (iii) The **electrodes** are the two conductors carrying current to the electrolyte. The positive-connected electrode is called the **anode** and the negative-connected electrode the **cathode**.

   (iv) When two copper wires connected to a battery are placed in a beaker containing a salt water solution, then current will flow through the solution. Air bubbles appear around the wires as the water is changed into hydrogen and oxygen by electrolysis.

3 **Electroplating** uses the principle of electrolysis to apply a thin coat of one metal to another metal. Some practical applications include the tin-plating of steel, silver-plating of nickel alloys and chromium-plating of steel. If two copper electrodes connected to a battery are placed in a beaker containing copper sulphate as the electrolyte it is found that the cathode (i.e. the electrode connected to the negative terminal of the battery) gains copper whilst the anode loses copper.

4 The purpose of an **electric cell** is to convert chemical energy into electrical energy. A **simple cell** comprises two dissimilar conductors (electrodes) in an electrolyte. Such a cell is shown in *Fig 1*, comprising copper and zinc electrodes. An electric current is found to flow between the electrodes. Other possible electrode pairs

exist, including zinc-lead and zinc-iron. The electrode potential (i.e. the p.d. measured between the electrodes) varies for each pair of metals. By knowing the emf of each metal with respect to some standard electrode the emf of any pair of metals may be determined. The standard used is the hydrogen electrode. The **electrochemical series** is a way of listing elements in order of electrical potential, and *Table 1* shows a number of elements in such a series. In a simple cell two faults exist—those due to polarization and local action (see *Problem 1*).

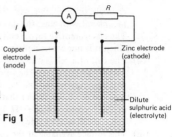

Fig 1

TABLE 1 PART OF THE ELECTROCHEMICAL SERIES

Potassium
sodium
aluminium
zinc
iron
lead
hydrogen
copper
silver
carbon

5 When two metals are used in a simple cell the electrochemical series may be used to predict the behaviour of the cell:
   (i) The metal that is higher in the series acts as the negative electrode, and vice-versa. For example, the zinc electrode in the cell shown in *Fig 1* is negative and the copper electrode is positive.
   (ii) The greater the separation in the series between the two metals the greater is the emf produced by the cell.
6 The electrochemical series is representative of the order of reactivity of the metals and their compounds:
   (i) The higher metals in the series react more readily with oxygen and vice-versa.
   (ii) When two metal electrodes are used in a simple cell the one that is higher in the series tends to dissolve in the electrolyte.
7 (i) **Corrosion** is the gradual destruction of a metal in a damp atmosphere by means of simple cell action. In addition to the presence of moisture and air required for rusting, an electrolyte, an anode and a cathode are required for corrosion. Thus, if metals widely spaced in the electrochemical series, are used in contact with each other in the presence of an electrolyte, corrosion will occur. For example, if a brass valve is fitted to a heating system made of steel, corrosion will occur.
   (ii) The **effects of corrosion** include the weakening of structures, the reduction of the life of components and materials, the wastage of materials and the expense of replacement.
   (iii) Corrosion may be **prevented** by coating with paint, grease, plastic coatings and

enamels, or by plating with tin or chromium. Also, iron may be galvanised, i.e., plated with zinc, the layer of zinc helping to prevent the iron from corroding.

8 (i) The **electromotive force (emf)**, E, of a cell is the p.d. between its terminals when it is not connected to a load (i.e., the cell is on 'no-load').

   (ii) The emf of a cell is measured by using a **high resistance voltmeter** connected in parallel with the cell. The voltmeter must have a high resistance otherwise it will pass current and the cell will not be on no-load. For example, if the resistance of a cell is 1 Ω and that of a voltmeter 1 MΩ then the equivalent resistance of the circuit is 1 MΩ + 1Ω, i.e. approximately 1 MΩ, hence no current flows and the cell is not loaded.

   (iii) The voltage available at the terminals of a cell falls when a load is connected. This is caused by the **internal resistance** of the cell which is the opposition of the material of the cell to the flow of current. The internal resistance acts in series with other resistances in the circuit. *Fig 2* shows a cell of emf E volts and internal resistance r, XY represents the terminals of the cell. When a load (shown as resistance R) is not connected, no current flows and the terminal p.d., $V = E$. When R is connected a current I flows which causes a voltage drop in the cell, given by Ir. The p.d. available at the cell terminals is less than the emf of the cell and is given by: $V = E - Ir$.

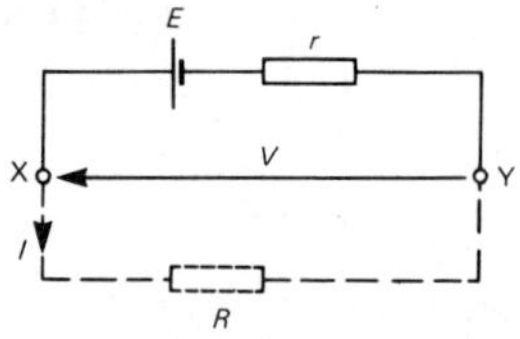

Fig 2

Thus if a battery of emf 12 volts and internal resistance 0.01 Ω delivers a current of 100 A, the terminal p.d., $V = 12 - (100)(0.01) = 12 - 1 = 11$ V.

   (iv) When a current is flowing in the direction shown in *Fig 2* the cell is said to be discharging ($E > V$).

   (v) When a current flows in the opposite direction to that shown in *Fig 2* the cell is said to be charging ($V > E$).

9 A **battery** is a combination of more than one cell. The cells in a battery may be connected in series or in parallel.

   (i) *For cells connected in series*:
   Total emf = sum of cell's emf's
   Total internal resistance = sum of cell's internal resistances

   (ii) *For cells connected in parallel*:
   If each cell has the same emf and internal resistance:
   Total emf = emf of one cell

   Total internal resistance of n cell = $\frac{1}{n}$ × internal resistance of one cell
   (See *Problems 2 to 5*)

10 There are two main types of cell—primary cells and secondary cells.

   (i) **Primary cells** cannot be recharged, that is, the conversion of chemical energy to electrical energy is irreversible and the cell cannot be used once the chemicals are exhausted. Examples of primary cells include the Leclanché cell and the mercury cell (see *Problems 6 and 7*). Practical applications of such cells include torch and transistor radio batteries and similar portable electrical equipment.

   (ii) **Secondary cells** can be recharged after use, that is, the conversion of chemical energy to electrical energy is reversible and the cell may be used many times. Examples of secondary cells include the lead-acid cell and alkaline cells (see *Problems 8 to 10*). Practical applications of such cells include car batteries,

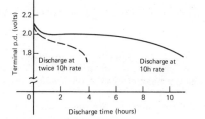

Fig 3

telephone circuits and for traction purposes—such as milk delivery vans and fork lift trucks.

11 The **capacity** of a cell is measured in ampere-hours (A h). A fully charged 50 A h battery rated for 10 h discharge can be discharged at a steady current of 5 A for 10 h, but if the load current is increased to 10 A then the battery is discharged in 3–4 h, since the higher the discharge current, the lower is the effective capacity of the battery. Typical discharge characteristics for a lead-acid cell are shown in *Fig 3*.

B. WORKED PROBLEMS ON THE CHEMICAL EFFECTS OF ELECTRICITY

Problem 1 (a) What is polarization, and how can the effect be overcome?
(b) Explain the term 'local action' when referred to a simple cell.

(a) (a) If the simple cell shown in *Fig 1* is left connected for some time, the current I decreases fairly rapidly. This is because of the formation of a film of hydrogen bubbles on the copper anode. This effect is known as the polarization of the cell. The hydrogen prevents full contact between the copper electrode and the electrolyte and this increases the internal resistance of the cell. The effect can be overcome by using a chemical depolarizing agent or depolarizer, such as potassium dichromate which removes the hydrogen bubbles as they form. This allows the cell to deliver a steady current.

(b) When commercial zinc is placed in dilute sulphuric acid, hydrogen gas is liberated from it and the zinc dissolves. The reason for this is that impurities, such as traces of iron, are present in the zinc which set up small primary cells with the zinc. These small cells are short-circuited by the electrolyte, with the result that localized currents flow causing corrosion. This action is known as local action of the cell. This may be prevented by rubbing a small amount of mercury on the zinc surface, which forms a protection layer on the surface of the electrode.

Problem 2 Eight cells, each with an internal resistance of 0.2 Ω and an emf of 2.2 V are connected (a) in series, (b) in parallel. Determine the emf and the internal resistance of the batteries so formed.

(a) When connected in series, total emf = sum of cell's emf = 2.2 × 8 = **17.6 V**

total internal resistance = sum of cell's internal resistance = 0.2 × 8 = **1.6 Ω**

141

(b) When connected in parallel, total emf = emf of one cell = **2.2 V**.

total internal resistance of 8 cells = $\frac{1}{8} \times$ internal resistance of one cell

$$= \frac{1}{8} \times 0.2 = \mathbf{0.025 \ \Omega}$$

Problem 3 A cell has an internal resistance of 0.02 Ω and an emf of 2.0 V. Calculate its terminal p.d. if it delivers (a) 5 A, (b) 50 A.

(a) Terminal p.d., $V = E - Ir$ where E = emf of cell, I = current flowing
and r = internal resistance of cell (see para. 8).
$E = 2.0$ V, $I = 5$ A and $r = 0.02$ Ω.
Hence $V = 2.0 - (5)(0.02) = 2.0 - 0.1 = \mathbf{1.9 \ V}$.
(b) When the current is 50 A, terminal p.d., $V = E - Ir = 2.0 - 50 (0.02)$
i.e., $V = 2.0 - 1.0 = \mathbf{1.0 \ V}$.
Thus the terminal p.d. decreases as the current drawn increases.

Problem 4 The p.d. at the terminals of a battery is 25 V when no load is connected and 24 V when a load taking 10 A is connected. Determine the internal resistance of the battery.

When no load is connected the emf of the battery, E, is equal to the terminal p.d., V, i.e., $E = 25$ V.
When current $I = 10$ A and terminal p.d. $V = 24$ V, then $V = E - Ir$
i.e., $24 = 25 - (10)r$.
Hence, rearranging, gives $10r = 25 - 24 = 1$

and the internal resistance, $r = \dfrac{1}{10} = \mathbf{0.1 \ \Omega}$.

Problem 5 Ten 1.5 V cells, each having an internal resistance of 0.2 Ω, are connected in series to a load of 58 Ω. Determine (a) the current flowing in the circuit and (b) the p.d. at the battery terminals.

(a) For ten cells, battery emf, $E = 10 \times 1.5 = 15$ V,
and the total internal resistance,

$r = 10 \times 0.2 = 2$ Ω.

When connected to a 58 Ω load the circuit is as shown in *Fig 4*.

Current $I = \dfrac{\text{emf}}{\text{total resistance}} = \dfrac{15}{58 + 2} = \dfrac{15}{60} = \mathbf{0.25 \ A}$.

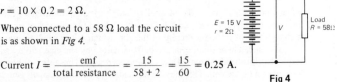

Fig 4

(b) P.d. to battery terminals, $V = E - Ir$ (see para. 8)
i.e. $V = 15 - (0.25)(2) = \mathbf{14.5 \ V}$.

Problem 6 Make a fully labelled sketch of a dry Leclanché cell and state typical applications of such a cell.

A typical dry Leclanché cell is shown in *Fig 5*. Such a cell has an emf of about 1.5 V when new, but this falls rapidly if in continuous use due to polarization (see *Problem 1*). The hydrogen film on the carbon electrode forms faster than can be

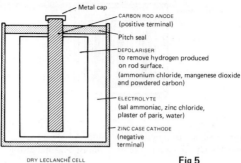

DRY LECLANCHÉ CELL **Fig 5**

dissipated by the depolarizer. The Leclanché cell is suitable only for intermittent use, applications including torches, transistor radios, bells, indicator circuits, gas lighters, controlling switch-gear, and so on. The cell is the most commonly used of primary cells, is cheap, requires little maintenance and has a shelf life of about 2 years.

Problem 7 Make a fully labelled sketch of a mercury cell and state applications where such a cell may be used in preference to a Leclanché cell.

A typical mercury cell is shown in *Fig 6*. Such a cell has an emf of about 1.3 V which remains constant for a relatively long time. Its main advantages over the Leclanché cell is its smaller size and its long shelf life. Typical practical applications include hearing aids, medical electronics and for guided missiles.

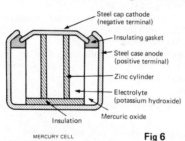

MERCURY CELL **Fig 6**

Problem 8 Describe the construction of a typical lead-acid cell and state the sequence for the discharge and charge of such a cell.

A typical lead-acid cell is constructed of:
(i) A **container** made of glass, ebonite or plastic.
(ii) **Lead plates**
    (a) the negative plate (cathode) consists of spongy lead,
    (b) the positive plate (anode) is formed by pressing lead peroxide into the lead grid.
The plates are interleaved as shown in the plan view of *Fig 7* to increase their effective cross-sectional area and to minimise internal resistance.

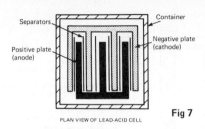

Fig 7

PLAN VIEW OF LEAD-ACID CELL

(iii) **Separators** made of glass, celluloid or wood.
(iv) An **electrolyte** which is a mixture of sulphuric acid and distilled water.

The relative density (or specific gravity) of a lead-acid cell, which may be measured using a hydrometer, varies between about 1.26 when the cell is fully charged to about 1.19 when discharged. The terminal p.d. of a lead-acid cell is about 2 V.

When a cell supplies current to a load it is said to be **discharging**. During discharge:
(i) the lead peroxide (positive plate) and the spongy lead (negative plate) are converted into the lead sulphate, and
(ii) the oxygen in the lead peroxide combines with hydrogen in the electrolyte to form water. The electrolyte is therefore weakened and the relative density falls.

The terminal p.d. of a lead-acid cell when fully discharged is about 1.8 V.

A cell is **charged** by connecting a d.c. supply to its terminals, the positive terminal of the cell being connected to the positive terminal of the supply. The charging current flows in the reverse direction to the discharge current and the chemical action is reversed. During charging:
(i) the lead sulphate on the positive and negative plates is converted back to lead peroxide and lead respectively, and
(ii) the water content of the electrolyte decreases as the oxygen released from the electrolyte combines with the lead of the positive plate. The relative density of the electrolyte thus increases.

The colour of the positive plate when fully charged is dark brown and when discharged is light brown. The colour of the negative plate when fully charged is grey and when discharged is light grey.

Problem 9 Briefly describe the construction of an alkaline cell.

There are two main types of alkaline cell—the nickel-iron cell and the nickel-cadmium cell. In both types the positive plate is made of nickel hydroxide enclosed in finely perforated steel tubes, the resistance being reduced by the addition of pure nickel or graphite. The tubes are assembled into nickel-steel plates.

In the nickel-iron cell, (sometimes called the Edison cell or nife cell), the negative plate is made of iron oxide, with the resistance being reduced by a little mercuric oxide, the whole being enclosed in perforated steel tubes and assembled in steel plates. In the nickel-cadmium cell the negative plate is made of cadmium. The electrolyte in each type of cell is a solution of potassium hydroxide which does not undergo any chemical change and thus the quantity can be reduced to a minimum. The plates are separated by insulating rods and assembled in steel containers which are then enclosed in a non-metallic crate to insulate the cells from one another. The average discharge p.d. of an alkaline cell is about 1.2 V.

Problem 10 Compare the advantages and disadvantages of an alkaline cell over a lead-acid cell and state where an alkaline cell may be used.

Advantages of an alkaline cell (for example, a nickel-cadmium cell or a nickel-iron cell) over a lead-acid cell include:
(i) More robust construction;
(ii) Capable of withstanding heavy charging and discharging currents without damage;
(iii) Has a longer life;
(iv) For a given capacity is lighter in weight;
(v) Can be left indefinitely in any state of charge or discharge without damage;
(vi) Is not self-discharging.

Disadvantages of an alkaline cell over a lead-acid cell include:
(i) Is relatively more expensive;
(ii) Requires more cells for a given emf;
(iii) Has a higher internal resistance;
(iv) Must be kept sealed;
(v) Has a lower efficiency.

Alkaline cells may be used in extremes of temperature, in conditions where vibration is experienced or where duties require long idle periods or heavy discharge currents. Practical examples include traction and marine work, lighting in railway carriages, military portable radios and for starting diesel and petrol engines. However, the lead acid cell is the most common one in practical use.

C. WORKED PROBLEMS ON THE CHEMICAL EFFECTS OF ELECTRICITY

(a) SHORT ANSWER PROBLEMS

1 What is electrolysis?

2 What is an electrolyte?

3 Conduction in electrolytes is due to ....

4 A positive connected electrode is called the .... and the negative connected electrode the ....

5 Name two practical applications of electrolysis.

6 The purpose of an electric cell is to convert .... to ....

7 Make a labelled sketch of a simple cell.

8 What is the electrochemical series?

9 What is corrosion?

10 Name two effects of corrosion and state how they may be prevented.

11 What is meant by the emf of a cell? How may the emf of a cell be measured?

12 Define internal resistance.

13 If a cell has an emf of E volts, an internal resistance of r ohms and supplies a current I amperes to a load, the terminal p.d. V volts is given by: $V = \ldots\ldots\ldots\ldots$

14 Name the two main types of cells.

15 Explain briefly the difference between primary and secondary cells.

16 Name two types of primary cells.

17 Name two types of secondary cells.

18 State three typical applications of primary cells.

19 State three typical applications of secondary cells.

20 In what units are the capacity of a cell measured?

(b) MULTI-CHOICE PROBLEMS (answers on page 149)

1 The terminal p.d. of a cell of emf 2 V and internal resistance 0.1 Ω when supplying a current of 5 A will be:
(a) 1.5 V; (b) 2 V; (c) 1.9 V; (d) 2.5 V.

2 Five cells, each with an emf of 2 V and internal resistance 0.5 Ω are connected in series. The resulting battery will have:
(a) an emf of 2 V and an internal resistance of 0.5 Ω;
(b) an emf of 10 V and an internal resistance of 2.5 Ω;
(c) an emf of 2 V and an internal resistance of 0.1 Ω;
(d) an emf of 10 V and an internal resistance of 0.1 Ω.

3 If the five cells of *Problem 2* are connected in parallel the resulting battery will have:
(a) an emf of 2 V and an internal resistance of 0.5 Ω;
(b) an emf of 10 V and an internal resistance of 2.5 Ω;
(c) an emf of 2 V and an internal resistance of 0.1 Ω;
(d) an emf of 10 V and an internal resistance of 0.1 Ω.

4 Which of the following statements is false?
(a) A Leclanché cell is suitable for use in torches;
(b) A nickel-cadmium cell is an example of a primary cell;
(c) When a cell is being charged its terminal p.d. exceeds the cell emf;
(d) A secondary cell may be recharged after use.

5 Which of the following statements is false?
When two metal electrodes are used in a simple cell, the one that is higher in the electrochemical series:
(a) tends to dissolve in the electrolyte;
(b) is always the negative electrode;
(c) reacts most readily with oxygen;
(d) acts as the anode.

6 Five 2 V cells, each having an internal resistance of 0.2 Ω, are connected in series to a load of resistance 14 Ω. The current flowing in the circuit is:
(a) 10 A; (b) 1.4 A; (c) 1.5 A; (d) $\frac{2}{3}$ A.

7 For the circuit of *Problem 6*, the p.d. at the battery terminals is:
(a) 10 V; (b) $9\frac{1}{3}$ V; (c) 0 V; (d) $10\frac{2}{3}$ V.

8 Which of the following statements is true?
   (a) The capacity of a cell is measured in volts;
   (b) A primary cell converts electrical energy into chemical energy;
   (c) Galvanising iron helps to prevent corrosion;
   (d) A positive electrode is termed the cathode.

(c) CONVENTIONAL PROBLEMS

1 Twelve cells, each with an internal resistance of 0.24 Ω and an emf of 1.5 V are connected (a) in series, (b) in parallel. Determine the emf and internal resistance of the batteries so formed. [(a) 18 V, 2.88 Ω; (b) 1.5 V, 0.02 Ω]

2 A piece of chromium and a piece of iron are placed in an electrolyte in a container. A d.c. supply is connected between the pieces of metal, the positive terminal being connected to the chromium. Explain what is likely to happen and why this happens.

3 With reference to conduction in electrolytes, explain briefly how silver plating of nickel alloys is achieved.

4 A cell has an internal resistance of 0.03 Ω and an emf of 2.2 V. Calculate its terminal p.d. if it delivers (a) 1 A, (b) 20 A, (c) 50 A.
[(a) 2.17 V; (b) 1.6 V; (c) 0.7 V]

5 The p.d. at the terminals of a battery is 16 V when no load is connected and 14 V when a load taking 8 A is connected. Determine the internal resistance of the battery. [0.25 Ω]

6 A battery of emf 20 V and internal resistance 0.2 Ω supplies a load taking 10 A. Determine the p.d. at the battery terminals and the resistance of the load.
[18 V; 1.8 Ω]

7 Ten 2.2 V cells, each having an internal resistance of 0.1 Ω are connected in series to a load of 21 Ω. Determine (a) the current flowing in the circuit, and (b) the p.d. at the battery terminals. [(a) 1 A; (b) 21 V]

8 Define electrolysis and use an example of electrolysis to show the meaning of the terms (a) electrode, (b) electrolyte.

9 The simple cell has two main faults—polarization and local action. Explain these two phenomena.

10 What is corrosion? State its effects and how it may be prevented.

11 Explain the difference between primary and secondary cells. Make a fully labelled sketch of one of each type of cell.

12 Describe the charging and discharging sequence of a simple lead-acid secondary cell.

13 Compare the performances and uses of the Leclanché dry cell and the mercury cell.

14 State three practical applications for each of the following cells:
(a) Dry Leclanché; (b) Mercury; (c) Lead-acid; (d) Alkaline.

15 How may the electrochemical series be used to predict the behaviour of a cell? Also, how is the series representative of the order of reactivity of the metals and their compounds?

Answers to multi-choice problems

CHAPTER 1 (page 5)

1 (c); 2 (d), 3 (b); 4 (c); 5 (b).

CHAPTER 2 (page 11)

1 (b); 2 (c); 3 (c); 4 (b), 5 (d).

CHAPTER 3 (page 18)

1 (d); 2 (c); 3 (a); 4 (b); 5 (a); 6 (b); 7 (b); 8 (c).

CHAPTER 4 (page 24)

1 (d); 2 (c); 3 (b); 4 (c), 5 (b); 6 (d); 7 (a); 8 (d).

CHAPTER 5 (page 35)

1 (d), 2 (b); 3 (c); 4 (a); 5 (c); 6 (a); 7 (d); 8 (a); 9 (b); 10 (b).

CHAPTER 6 (page 44)

1 (b); 2 (d); 3 (a); 4 (a); 5 (c); 6 (d); 7 (b).

CHAPTER 7 (page 52)

1 (c); 2 (g); 3 (d); 4 (c); 5 (e); 6 (b); 7 (a); 8 (a); 9 (i), 10 (e).

CHAPTER 8 (page 60)

1 (j); 2 (a), 3 (b); 4 (d); 5 (f); 6 (i); 7 (g); 8 (d); 9 (c); 10 (a).

CHAPTER 9 (page 66)

1 (f); 2 (e); 3 (i); 4 (c); 5 (h); 6 (b); 7 (d); 8 (a).

CHAPTER 10 (page 73)

1 (c); 2 (b); 3 (a); 4 (a); 5 (c).

CHAPTER 11 (page 81)

1 (c); 2 (d); 3 (b); 4 (a); 5 (d).

CHAPTER 12 (page 89)

1 (b); 2 (c); 3 (c); 4 (a); 5 (d); 6 (c); 7 (a); 8 (d).

CHAPTER 13 (page 102)

1 (d); 2 (b); 3 (a); 4 (c); 5 (d); 6 (b); 7 (b); 8 (c); 9 (c); 10 (d)

CHAPTER 14 (page 118)

1 (a), 2 (c); 3 (c); 4 (c); 5 (a); 6 (b); 7 (d); 8 (d); 9 (b); 10 (c).

CHAPTER 15 (page 134)

1 (d); 2 (b); 3 (c); 4 (c); 5 (c); 6 (c); 7 (d); 8 (a); 9 (c); 10 (b).

CHAPTER 16 (page 146)

1 (a); 2 (b); 3 (c); 4 (b); 5 (d); 6 (d); 7 (b); 8 (c).

Index

Absolute pressure, 39
Acceleration, 55
Acids, 21, 23
Air, composition of, 20
Alkali, 21, 23
Alkaline cell, 140, 144
Alloy, 15
Ammeter, 106
Ampere, 106
Angle of incidence, 74
    reflection, 74
Anode, 138
Atmospheric pressure, 39
Atom, 13, 16, 105

Barometer, 42
Base, 21
Battery, 140
Bourdon pressure gauge, 43
Brittleness, 8

Camera, 77
Capacity of cell, 141
Cathode, 138
Cells, 138
Celsius scale, 92
Centre of gravity, 27
Chemical effects of electricity, 138
Chemical equation, 21
    reactions, 20
Coefficient of friction, 63
Combustion, 20
Compound, 13
Compressive force, 7
Concave lens, 75, 77
Conduction, 94, 99
Conductors, 106, 108
Conservation of energy, 83
Convection, 94, 99
Convex lens, 75, 77
Corrosion, 139
CRO, 106
Coulomb, 106
Crystal, 14
Crystallisation, 14
Current, 106

Density, 2
Distance-time graphs, 46
Ductility, 8
Dynamic friction, 63

Effects of elastic current, 110
Efficiency, 84
Elasticity, 8
Elastic limit, 8
Electric bell, 129
Electric circuits, 105
    symbols, 105
Electrochemical series, 139
Electrode, 138
Electrolysis, 138
Electrolyte, 138
Electromagnet, 124
Electromagnetic induction, 125, 132
Electromagnetism, 122
Electromotive force, 140
Electrons, 105, 138
Electroplating, 138
Element, 13, 105
Energy, 83, 84, 110
Equilibrium, 27

Fleming's left-hand rule, 124
Fluid, 38
Focal length, 77
Focal point, 77
Force, 7, 27
Force-distance graph, 83
Fortin barometer, 43
Free fall, 55
Frequency, 70
Friction, 63
Fuel, 83
Fuse, 110

Gauge pressure, 39
Generator, a.c., 125, 132
Gravitational force, 56
Grip rule, 124

150

Heat, 92
Hooke's law, 8
Hydrometer, 2

Images, 77
Indicator, 21
Insulation, 100
Insulators, 106, 108
Internal resistance, 140
Ions, 21, 138

Joule, 83

Kelvin, 92
Kilowatt-hour, 110

Lamina, 27
Lamps in series and parallel, 109
Latent heat of fusion, 94
    vaporisation, 94
Lead acid cell, 140, 143
Leclanchè cell, 140, 142
Lenses, 75, 77
Lifting magnet, 130
Light rays, 74
Liquid-in-glass thermometer, 92, 95
Local action, 141
Longitudinal waves, 69

Magnetic field, 122
    flux, 122
    force, 122
Magnifying glass, 77, 79
Malleability, 8
Manometer, 41
Mercury cell, 140, 143
Microscope, 79
Mixture, 14
Molecule, 13
Moments, 28
    principle of, 28
Motor, electric, 124, 130
Moving coil instrument, 124, 131
Multimeter, 106

Newton, 7, 56
Normal force, 63
Nucleus, 105

Ohm, 106

Ohmmeter, 106
Ohm's law, 107
Oxidation, 20
Oxides, 20

Parallel connection, 106, 108
Parallelogram of forces, 29
Pascal, 38
Periscope, 78
Permanent magnet, 122
Photocopier, 77
pH scale, 21
Pitch, 71
Plasticity, 8
Polarization, 141
Polycrystalline substances, 15
Potential difference, 106
Power, 84, 110
Pressure in fluids, 38
Primary cell, 140
Principal axis, 75
Principle of conservation of:
    energy, 83
    moments, 28
Projectors, 77, 80
Protons, 105
Pyrometers, 92

Radiation, 94, 100
Rays, 74
Real image, 77
Reflection, 70, 74
Refraction, 70, 74
Refrigerator, 99
Relative density, 2
Relay, 129
Resistance, 106
    thermometer, 92
    variation, 121
Resistivity, 121
Resultant, 29
Rusting, 20

Salt, 21
Scalar quantities, 27
Screw rule, 123, 124
Secondary cell, 140
Sensible heat, 94
Series connection, 106, 108
Shear force, 8
SI units, 1
Solenoids, 124
Solubility, 14

Solute, 14
Solution, 14
  solid, 15
Solvent, 14
Sound waves, 70
Specific heat capacity, 92
Spectacles, 77
Speed, 46
Spotlight, 77
Static friction, 63
Structure of matter, 13
Suspension, 14

Temperature, 92
  coefficient of resistance, 122
Tensile force, 7
  test, 8
Thermal movement, 101
Thermocouple, 92, 95

Thermodynamic scale, 92
Thermometers, 92
Transverse waves, 69
Triangle of forces, 29

Vacuum flask, 100
Vector addition, 29
  quantities, 27
Velocity, 47
Velocity-time graph, 47
Virtual image, 77
Volt, 106

Watt, 84, 110
Wavelength, 69
Waves, 69
  velocity of, 70
Work done, 83